CERAMICS GLAZE TECHNOLOGY

Pergamon Titles of Related Interest

BYE	Portland Cement
FORD	Ceramics Drying
INSTITUTE OF CERAMICS	Health and Safety in Ceramics, 2nd Edition
MASKALL & WHITE	Vitreous Enamelling
RYAN	Properties of Ceramic Raw Materials, 2nd Edition
WORRALL	Ceramic Raw Materials, 2nd Edition

Pergamon Related Journals *(Sample copies gladly sent on request)*

Journal of the Australian Ceramic Society

International Journal of Solids and Structures

Journal of Physics and Chemistry of Solids

Journal of the Mechanics and Physics of Solids

CERAMICS GLAZE TECHNOLOGY

J. R. TAYLOR
and
A. C. BULL

Cookson Ceramics & Antimony Ltd
Stoke-on-Trent, UK

Published on behalf of
THE INSTITUTE OF CERAMICS
By

PERGAMON PRESS

OXFORD . NEW YORK . TORONTO . SYDNEY . FRANKFURT
TOKYO . SAO PAOLO . BEIJING

U.K.	Pergamon Press, Headington Hill Hall, Oxford OX3 0BW, England
U.S.A.	Pergamon Press Inc., Maxwell House, Fairview Park, Elmsford, New York 10523, U.S.A.
CANADA	Pergamon Press Canada, Suite 104, 150 Consumers Road, Willowdale, Ontario M2J 1P9, Canada
AUSTRALIA	Pergamon Press (Aust.) Pty. Ltd., P.O. Box 544, Potts Point, N.S.W. 2011, Australia
FEDERAL REPUBLIC OF GERMANY	Pergamon Press GmbH, Hammerweg 6, D-6242 Kronberg, Federal Republic of Germany
JAPAN	Pergamon Press, 8th Floor, Matsuoka Central Building, 1-7-1 Nishishinjuku, Shinjuku-ku, Tokyo 160, Japan
BRAZIL	Pergamon Editora Ltda., Rue Eça de Queiros, 346, CEP 04011, São Paulo, Brazil
PEOPLE'S REPUBLIC OF CHINA	Pergamon Press, Qianmen Hotel, Beijing, People's Republic of China

First edition 1986

Library of Congress Cataloging-in-Publication Data
Taylor, J. R.
Ceramics glaze technology.
Includes bibliographical references.
1. Glazing (Ceramics) 2. Glazes. I. Bull, A. C.
II. Title.
TP812.T39 1986 666'.444 85-29700

British Library Cataloguing in Publication Data
Taylor, J. R.
Ceramics glaze technology.
1. Glazing (Ceramics)
I. Title II. Bull, A.C.
738.1'2 TP812
ISBN 0-08-033465-2 Hard cover
ISBN 0-08-033466-0 Flexicover

Printed in Great Britain by A. Wheaton & Co. Ltd, Exeter

CONTENTS

INTRODUCTION

For thousands of years, man has made use of natural mineral, vegetable and animal resources to provide his shelter, food, tools, weapons and a variety of other functional items. For example, flints were used to make tools and weapons and wood, straw and rocks have been employed as building materials. Mineral deposits, including clays, have been a source of construction materials and also of raw material for the fabrication of everyday household articles.

The clay-containing deposits originally found suitable for making household goods were capable of being readily shaped into articles which, after drying in the sun, could be used to store dry foods such as grain. However, unfired clay vessels, although possessing some strength, are easily broken, are porous and suffer a severe loss of integrity on contact with water. At a later stage of history we assume it was noticed that subjecting the clay articles to the high temperature of a household fire produced a significant increase in the mechanical strength of the vessel and a reduction in its porosity. Although this treatment could give the fired clay articles a longer useful life than their sun-dried predecessors, they would still be porous and liable to staining from food residues. These fired ceramics thus were not ideal articles from which food could be consumed or in which liquids could be stored.

Clearly vessels having an impervious coating which could be cleaned easily and which were capable of withstanding moderately high temperatures were desirable for domestic use. We can surmise that some such ceramic articles were produced, by chance perhaps, when certain compounds were present either in the fuels or the raw materials employed in the making of fired ceramic articles so that an impervious glassy coating or "glaze" was produced as an outer skin on the ware. The advantages offered by such glazed wares were obvious and as a result of their more widespread use, a greater understanding of the nature of glazes was gained. The decorative potential of ceramic glazes, in addition to these functional attributes, was also appreciated.

The understanding of the nature of ceramic glazes has resulted in the ability of the technologist to design glazes with specific physical and chemical properties to suit particular requirements. The relative importance of different properties in the fired glaze might alter as requirements are modified but a glaze can be considered to be a usually smooth, thin (75–500 μm) and essentially glassy coating on a ceramic body, this coating adhering strongly to the underlying substrate.

To understand the properties of glazes, a knowledge of the properties of glasses is helpful.[1,2] For a suitable glaze to be designed, the characteristics of the substrate need to be taken into account to ensure the compatibility of the glaze and substrate[3] so that the required quality of the glazed product is attained after firing and so that the glaze develops the right surface finish.[4] Attention needs to be paid to the necessity for the fired glaze to be unaffected by water and have good durability to acid and alkaline solutions. The glaze will be expected to be resistant to physical wear to a degree depending on the nature of the use to which the glazed article is put and other properties such as for example scratch-slip or metal marking resistance might have to be considered.

Up to this point, the term "glaze" has been associated with a fired glassy coating on a ceramic substrate, but the term is also applied to the unfired composition once the essential inorganic component parts have been mixed together. As the selection made of these constituent materials (which may be preformed glasses, minerals or manufactured chemicals) can affect both the ease of processing of the raw glaze mix and properties of the fired glaze, an understanding of *their* properties and the glaze production processes is clearly desirable. The use of minor amounts of various additives offers many benefits because they can assist in the sundry manufacturing processes which occur prior to the glost firing and the glaze technologist needs to appreciate the effects of using materials to achieve the best results.

This volume will deal with these several and various aspects of glazes. After an initial discussion of the nature and properties of glasses, the raw materials, production, glazing and firing processes associated with glazes will be described, together with the testing methods employed in the various areas of glaze manufacture and use. Mention will also be made of some common faults associated with glaze whether or not these are due to an unwanted characteristic of the glaze composition, its processing or its use.

REFERENCES

1. Moore, H. *Trans. Brit. Ceram. Soc.* **55**, 589–600 (1956).
2. Bloor, E. C. *Trans. Brit. Ceram. Soc.* **55,** 631–660 (1956).
3. Norris, A. W. *Trans. Brit. Ceram. Soc.* **55,** 674–688 (1956).
4. James, W. and Norris, A. W. *Trans. Brit. Ceram. Soc.* **55,** 601–630 (1956).

Chapter 1
Definitions

1.1 DEFINITION OF GLAZE

We can examine the definition from four distinct viewpoints showing how an interpretation differs according to what is required of the word. In the first authority a glaze is defined[1] as *the vitreous composition used for glazing pottery, etc.* A technical dictionary[2] allots the meaning to glaze as being *a brilliant glasslike surface given to tiles, bricks, etc.* This is notable in that no specific mention is made of tableware or sanitaryware. An early ceramic treatise[3] asserts that glazes are *vitreous coatings with which we cover bodies either to decorate them or to make them impermeable and we may consider them as glass in the widest meaning of that word.* However, as modern ceramists, we might be guided by an arcane definition[4] which states that a glaze is *the thin glassy layer formed on the surface of a ceramic product by firing in the presence of an alkali vapour.*

The term glaze is also applied to the prepared mixture of materials, which is either a powder or a suspension in water, ready for application to ceramic ware by dipping or spraying. After suitable heat treatment this powdered mixture vitrifies and develops specific properties appropriate to the designed use of the glaze on the ceramic body. The composition of the glaze is chosen to ensure certain well-defined properties, such as adhesion to the substrate, correspondence of thermal expansion, transparency or opacity, surface texture and resistance to chemical attack, are possessed by the glasslike coating.

It is the purpose of this book to expand these statements and explain in detail their relationship to glaze technology.

1.2 DEFINITION OF THE TERM GLASS

The common factor in the formal definition is the need to use the nature of glass to describe the essence of glaze, but unless we can understand what is meant by "glass" no progress can be made. Bourry[5] recognised a glass by such characteristic properties as brittleness, hardness, gloss and transparency, but as scientists and technologists we need a more precise definition as to its nature.

Many dictionaries have used as a basis the definition suggested by the American Society for Testing Materials[6] in 1945. This states that *glass is an inorganic product of fusion which has cooled to a rigid condition without crystallising.* The British Standards Institution[7] adopted the same phraseology in 1962. Whilst there is no mention of colour or opacity, it is easy to relate the material so defined to our everyday idea of glass and by extension to the material we see as the impervious layer on ceramic substrates.

Such a definition is no longer adequate. Materials which have the properties of glass can be made by unconventional routes unrelated to cooling a molten mass. Glasslike materials can be prepared by shock wave vitrification of crystals,[8] by sputtering and splat cooling,[9] by deposition from the vapour phase,[10] by flame hydrolysis,[11] by the sol-gel process,[12–14] by laser pulse heating[16] and from some organic compounds. Almost any substance can perhaps be prepared in the glassy state if cooled sufficiently fast.

These novel methods led to an attempt at a new definition, *glass is a non-crystalline solid.*[15] A contemporary alternative suggested that glass should be defined as a solid with dense packing of the atoms in which crystalline arrangements do not exist beyond a few interatomic distances. The existence of the alternative methods produced a redefinition proposed by the United States National Academy for Science Research Council. This describes *glass* as *an X-ray amorphous material which exhibits the glass transition, this being defined as that phenomenon in which a solid amorphous phase exhibits with changing temperature a more or less sudden change in the derivative thermodynamic properties such as heat capacity and expansion coefficient from crystal-like to liquid-like values. The temperature of the transition is called the glass (transition) temperature and denoted by* T_g

Preparation of glasses below the transformation temperature would provide a possibility of obtaining glasses which cannot be obtained at higher temperatures because of a tendency to crystallise. The relevance of these characteristic features to glaze technology will be examined in subsequent sections.

1.3 THE GLASSY STATE

Consider a compound in the molten state. A change of state occurs at a temperature when the liquid melt becomes solid. If the material is a pure crystal there is a volume change and heat is evolved. The two phases, liquid and solid, can be distinguished by their different mechanical properties. The liquid is easily deformed whilst the solid has mechanical strength and elasticity. This difference can be explained by the fixed, ordered position which atoms or molecules have in a crystal and their free mobility in a liquid. In a crystalline solid there is both short-range and long-range ordering of the molecules. There can be only a degree of short-range ordering of molecules in a liquid.

Consider the relation between volume change and temperature in a cooling liquid (Fig. 1.1). In certain conditions the phase change from liquid to solid does not occur at a definite temperature. There are examples of liquids which solidify gradually on cooling. In such examples the solid does not liquefy promptly as the temperature rises and a definite phase change is not discernible. At a high enough temperature the material is liquid and at a low enough temperature the material is solid. There is no recognisable discontinuous volume change and no heat is evolved. Between the two phase conditions there is a transition region in which the viscosity increases as the temperature falls and it is impossible to say whether the material should be classed as a liquid or as a solid. At one chosen temperature the material might be in a state which resists an instanteously applied force, yet it would deform (even if only slightly) if the same force was applied continuously. The material is a liquid of very high viscosity. Further reduction in temperature increases the viscosity further and the mass develops the properties of an elastic solid. A material which behaves in this way is "glass". This can be summarised in the form "glass is a state of matter which maintains the energy, volume and atomic arrangement of a liquid but for which the changes in energy and volume with temperature are similar in magnitude to those of a crystalline solid".[17] The X-ray diffraction pattern of this glassy material shows only diffuse haloes which are virtually indistinguishable from the pattern of a liquid, compared with a sharp arrangement of lines with a crystal. There is, therefore a lack of long-range ordering within a glass. More detail of this aspect of glass behaviour and of the importance of this transition region of the change of state to the design of glaze–substrate fit will be examined later.

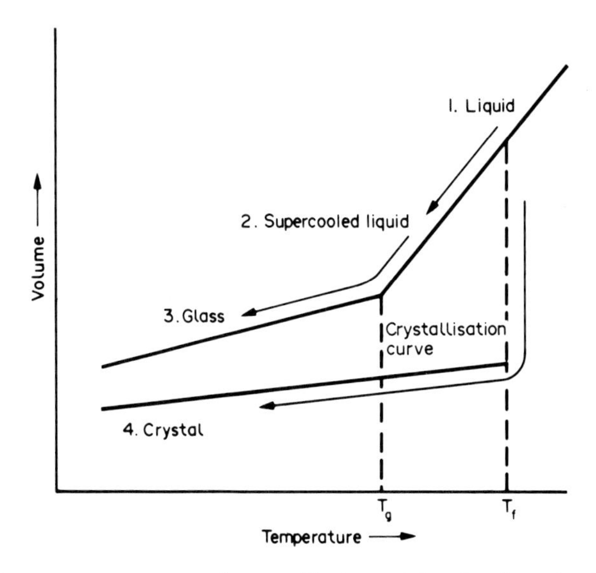

Fig. 1.1. Volume–temperature curves for crystallisation and for glass formation from liquid. T_f melting temperature, T_g transformation temperature.

1.4 GLASS STRUCTURE

Around 1925 our knowledge of the structure of glass was limited to the concept that glass is an inorganic non-crystalline isotropic solid. Some investigators regarded it either as a liquid of immense viscosity or as a supercooled liquid. Prior to the X-ray diffraction studies by Warren[18] and his co-workers, experiments designed to understand the nature of glass were unrewarding. Results of the new X-ray work showed that a glass of a normal silicate type could be represented as irregularly grouped silica aggregates with alkali or basic oxides distributed throughout the system. The whole system was linked, or bridged, through oxygen atoms. In this way the structure of glass was shown to be built from the same units that build crystalline forms of silica.[19, 20] In silica crystals, each silicon atom is four-co-ordinated, that is linked with four oxygen atoms at the corner of a tetrahedron.

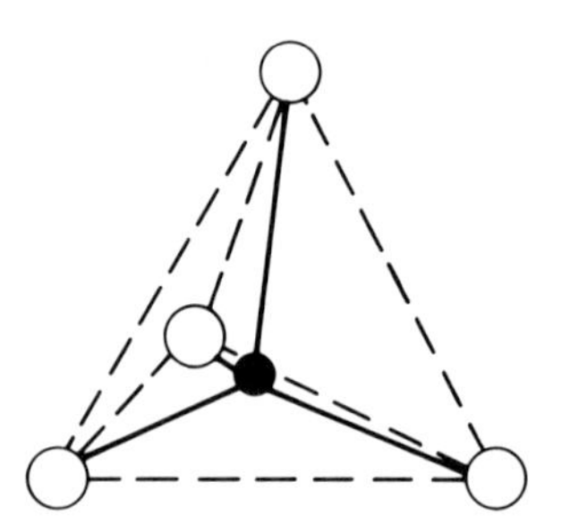

Fig. 1.2. Silicon-oxygen tetrahedron.

Each oxygen atom acts as a bridge to link with two silicon atoms in neighbouring tetrahedra. The arrangement in a glass, however, is irregular and this precludes the development of crystalline properties. In vitreous silica the orientation of adjacent silicon–oxygen tetrahedra is variable, but in crystalline silica the orientation is constant throughout.

The mean spacing of the tetrahedra in fused silica glass is slightly greater than the regular spacing in say tridymite. Nevertheless, within a short range the arrangement and spacing of the tetrahedra are regular. In ordinary silicate glasses the average spacing of the silica tetrahedra is less close than in pure silica glass to accommodate the alkali and basic oxide constituents. If the proportion of these constituents increases, then the mean spacing increases in proportion.

An examination of the characteristics of glass-forming oxides led to the development of the random network theory of glass structure.[21] The ability of an oxide to form a tetrahedral configuration is not the sole criterion of glass-forming ability. In beryllium oxide (see Table 1.1[22]) Be is surrounded by a tetrahedron of O cations but does not form a glass on its own. Nevertheless, following the assumption that the inter-atomic forces in glasses and crystals must be similar, it was deduced that the atoms should be linked in glasses in a three-dimensional network as they are linked in crystals. The network in glass could not have a long-range periodicity since glasses do not exhibit sharp X-ray diffraction patterns. Crystals have a regular lattice and sharp patterns and by corollary the bonds linking the structural units of the glass are sufficiently distorted to permit only short-range order in an otherwise random way. This is illustrated in Fig. 1.3 where a hypothetical oxide M_2O_3 is drawn, in two dimensions, as it might exist in a crystal phase and in a glass form.

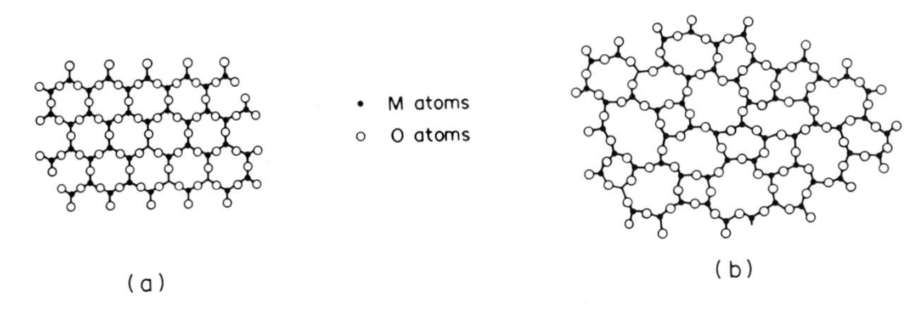

Fig. 1.3. Two-dimensional representation of the structure of (a) M_2O_3 as a crystal; (b) the glass form of M_2O_3.

The original picture proposed by Zacharaisen is still widely used. It gives the technologist an insight into some of the properties of glaze.[23-25] If a silicate glass does consist of a random network of tetrahedra linked through oxygen atoms, then the formation of a silica-based glass from its primary materials can be regarded as an opening in the oxygen links in the silica. When an oxygen link is disrupted, one side of the gap is without oxygen and the other side will have an oxygen with a valency bond unsatisfied. There is here an opportunity for a divalent metal oxide, say calcium oxide, CaO, to enter the gap, completing the bridge and retaining some rigidity in the structure. The original ≡Si–O–Si≡ link has been replaced by the succession ≡Si–O–Ca–O–Si≡.

An alternative event would be if an alkali metal oxide were to enter the gap. One Na atom joins the unsatisfied oxygen on one side while the remaining NaO group fully occupies the vacancy at the corner of the tetrahedra. These links are now seen to be ≡Si–O–Na and Na–O–Si≡. With all valencies satisfied in this way, the gap is not continuously bridged and there is now a break in the structure.

The formation of glass is impossible unless each SiO_4 tetrahedron is left with at least two corners unencumbered by alkali oxides. In other words, the limiting composition of a sodium oxide–silicon dioxide glass is Na_2O. SiO_2.[26] The three common alkali metal ions behave in similar ways. The lithium ion Li^+ is the smallest and the potassium ion K^+ is the largest of the three. In the network Li^+

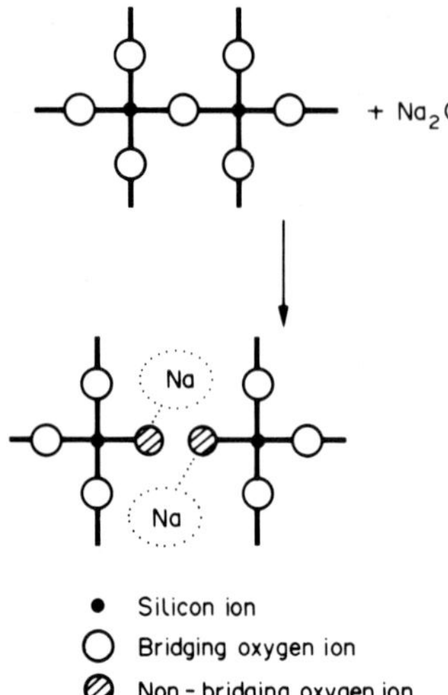

Fig. 1.4. Reaction between sodium oxide and silica tetrahedra. (In this figure, for simplicity, a two-dimensional representation of the SiO_4 groups is given; in the actual glass structure these groups take the form of tetrahedra.)

will tend to occupy smaller cells and K^+ will tend to occupy larger cells than are occupied by the intermediately sized sodium ion Na^+.

The strengthening effect of completing the bridge across the gap is manifest when the viscosity of a glass containing CaO in partial replacement for Na_2O shows a rapid increase with increase in CaO content. The solubility of the glass in water is also greatly reduced as the CaO enters the structure as a linkage.

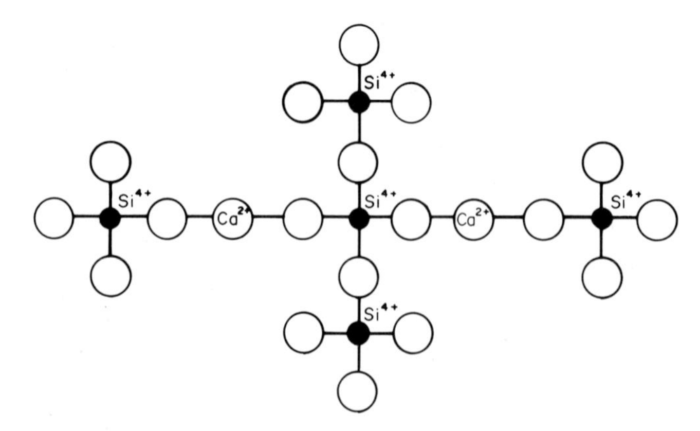

Fig. 1.5. Schematic representation of CaO in a silica network.

Only silicon dioxide based structures have been discussed thus far, but other oxides easily form glasses on their own. Boric oxide, B_2O_3, germanium dioxide, GeO_2, and phosphorus pentoxide, P_2O_5, are all known as "glass network formers" and with silica they provide the essential framework for mixed oxide glasses. SiO_2 and SiO_2–B_2O_3 systems provide the backbone of almost all commercial glazes.

Glasses can also be produced from arsenic trioxide, As_2O_3, and antimony trioxide, Sb_2O_3, if the melts are cooled rapidly. Other oxides are "conditional glass formers".[27] The oxides of aluminium, bismuth, gallium, molybdenum, selenium, tellurium, titanium and vanadium will not form glasses on their own, but will do so when melted with a second oxide. Except for aluminium and titanium, their usefulness in glaze is limited. An alternative description, sometimes used, is that these are "intermediate oxides".

Metal oxides (e.g. the alkali and alkaline earth oxides) which can modulate the properties of these glass-formers are called "network modifiers".

TABLE 1.1 IONIC FIELD STRENGTHS OF CATIONS PRESENT IN GLAZES OR FRITS

Ion	Ionic radius[28] (A)	Field* strength Z/r^2	Structural role in glaze or frit
B^{3+}	0.23	56.7	Network-forming ions
P^{5+}	0.35	40.8	
Si^{4+}	0.42	21.6	
Ge^{4+}	0.53	14.2	
Al^{3+}	0.51	11.5	Intermediate ions
Ti^{4+}	0.68	8.7	
Zr^{4+}	0.79	6.4	
Sn^{4+}	0.71	7.9	Network-modifying ions
Mg^{2+}	0.66	4.6	
Zn^{2+}	0.74	3.6	
Li^{+}	0.68	2.2	
Ca^{2+}	0.99	2.1	
Sr^{2+}	1.12	1.6	
Pb^{2+}	1.20	1.4	
Na^{+}	0.97	1.1	
Ba^{2+}	1.34	1.1	
K^{+}	1.33	0.6	

*Z=valency, r=ionic radius.

In crystals such as felspars, aluminium ions can be four-co-ordinated (tetrahedral) or in micas aluminium ions can be six-co-ordinated (octahedral) with oxygen. A tetrahedral group AlO_4 can replace a silica tetrahedra to assume the arrangement in Fig. 1.6.

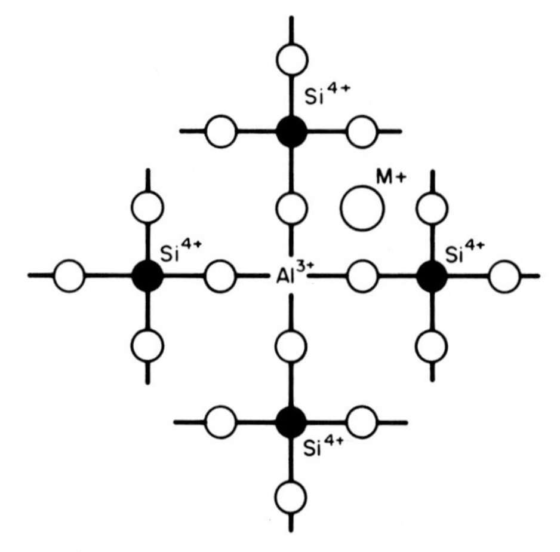

Fig. 1.6. Aluminium in a silicate network. (The structure is shown in a simplified form: the true structure is three-dimensional, the AlO_4 and SiO_4 groups having tetrahedral configurations.)

In a similar manner, with the introduction of Al_2O_3 as an intermediate oxide, or conditional glass-former, to a glass, one Al^{3+} ion will substitute for one Si^{4+} ion. In other words, an AlO_4 tetrahedron joins in the network formed by SiO_4 tetrahedra.[29] Electro-neutrality is maintained if there is an additional cation available and it can be located in an adjacent network "hole". This cation might be an alkali metal ion (see Fig. 1.6.) or even an alkaline earth ion if it can be shared between two neighbouring AlO_4 tetrahedra.

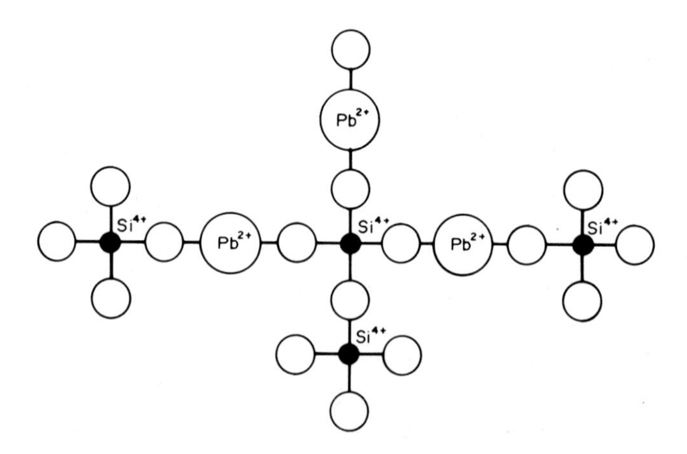

Fig. 1.7. Schematic representation of PbO in a silica network.

Lead oxide (PbO) and silica (SiO_2) form a very wide range of binary glasses, many with low melting characteristics. Lead can be thought to take part in the network formation. The Pb^{2+} ions act as links between two SiO_4 tetrahedra through bonds between corner oxygens (see Fig. 1.7.). An interpretation derived from this is that the network so formed is more open than one formed solely by SiO_4 tetrahedra.

With this simplistic picture of the structure of a glass, frit or glaze we can begin to visualise the physical properties which might be associated with a particular grouping of these elementary units. The viscosity at any particular temperature will depend upon the mean size of these groupings. The decrease in viscosity with increasing temperature indicates that the groups break down into smaller groups which would imply that rapid cooling from any higher temperature should give a frit of a lower specific gravity than would be obtained if the frit were allowed to cool slowly.

The very high viscosity of glass at ordinary temperatures implies that the mean molecular groupings are very large. There is consequently an inherent resistance to crystallisation.

As a consequence of the absence of crystallisation at lower temperatures, glass is sometimes referred to as a supercooled liquid. With an increasing degree of supercooling there would normally be an increasing tendency to crystallise, but the rapid rise in the already high viscosity of the melt means that the rearrangement of the structural units is unlikely. However, there are occasions when the firing cycle is modified and a glaze composition is designed to allow a limited degree of crystal formation to develop. When unwanted crystallisation occurs in a glaze, then knowledge of its nature can suggest methods of correcting the fault.

REFERENCES

1. *OED*, 1971.
2. *Dictionary of Science and Technology.* Chambers (1974).
3. Bourry, Emile. *Treatise on the Ceramic Industries* (translated by W. P. Rix), Scott Greenwood (1901).
4. Dodd, A. E. *Dictionary of Ceramics.* Newnes (1964).
5. Bourry, E. *Treatise on the Ceramic Industries* (translated by W. P. Rix), Scott Greenwood (1901).
6. ASTM C162–22.
7. *Glossary of Terms used in the Glass Industry,* BS 3447/1962, British Standards Institution, London.
8. De Carli, P. S. and Jamieson, J. C. "Formation of an amorphous form of quartz". *J. Chem. Phys.* **31,** 1675 (1959).
9. Klement, W., Willens, R. H. and Duwez, P. "Non-crystalline structure in solidified gold–silicon alloys". *Nature,* **187,** 869 (1960); Campbell, Grant and McKim, Richard. "Materials with a magnetic future". *New Scientist,* **94,** No. 1308, 637 (1982).
10. Kingery, W. D., Woulbroun, J. M. and Coble, R. L. "Glasses formed by vapour deposition". *Advances in Glass Technology,* Part 2. Plenum Press (1963); Keck, D. B. and Schultz, P. C. USP 3,711, 262 (1973).
11. Hyde, J. F. USP 2,272,342 (1942).
12. Roy, R. *J. Amer. Ceram. Soc.* **52,** 344 (1969).
13. Klein, Lisa C. Sol-gel Glass Technology *Glass Indust.* Jan. 1981, 14–17.
14. Mukherjee, S. P. *J. Non-cryst Solids,* **42,** 477–488 (1980).
15. Mackenzie, J. D. (Ed.) *Modern Aspects of the Vitreous State,* Vol. 3, p. 149, Butterworth (1964).

16. Nelson, A. and Blander, M. *J. Non-crystalline Solids,* **16,** 321 (1974).
17. Paul, A. *Chemistry of Glasses,* p. 4. Chapman & Hall (1982).
18. Warren, B. E. *J. Soc. Glass Tech.,* **24,** 159–165 (1940).
19. Paul, A. *Chemistry of Glasses,* p. 10. Chapman & Hall (1982).
20. Worrall, W. E. *Ceramic Raw Materials,* 2nd ed. p. 13. Pergamon (1982).
21. Zacharaisen, W. H. *J. Amer. Chem. Soc.,* **54,** 3841–3851 (1932).
22. *Table of Ionic Radii,* Handbook of Chemistry & Physics pp. F214–F215 CRC Press Inc. 1979.
23. Moore, H. *Trans. Brit. Ceram. Soc.,* **55,** 589–600 (1956).
24. Bloor, E. C. *Trans. Brit. Ceram. Soc.,* **55,** 631–660 (1956).
25. Ainsworth, L. *Trans. Brit. Ceram. Soc.,* **55,** 661–673 (1956).
26. McMillan, P. W. *Glass Ceramics,* 2nd ed. p. 19. Academic Press (1979).
27. Rawson, H. *Inorganic Glass Forming Systems,* pp. 8–9. Academic Press (1967).
28. *Handbook of Chemistry and Physics* (Eds. R. C. Weaste and M. J. Astle), pp. F214–F215. CRC Press Inc. (1979).
29. McMillan, P. W. *Glass Ceramics,* 2nd ed., p. 14. Academic Press (1979).

Chapter 2
Classification of Glaze

Within the definitions given for glaze there is an implicit acceptance that the coating can be applied to a variety of ceramic substrates having different design functions and properties. The requirements for each will differ widely. It will be expected, therefore, that glazes must be designed to fit a specification. For example, glazes for decorative ceramics can be unacceptable as glaze on tableware which comes into contact with food either in its preparation or storage. Accurate knowledge of what is expected from the glaze surface is essential before any recommendations can be made about its composition.

There are several classes into which the large number of possible glazes can be placed for identification:

i *Presence of specific element.* Glaze can be typified according to the presence or absence of the valuable flux lead oxide. Legal requirements and the need for safe working practices require unfired glaze containing fritted lead oxide to be marked accordingly and to have an acceptable level of resistance to acid attack in the working environment. Finished tableware coated with a lead glaze must also possess an acceptable level of resistance to chemical attack (Section 9.4). All practical compositions can be divided into either leadless or lead-containing ("low-solubility") glaze.

ii *Presence or absence of frit.* An alternative classification divides glaze according to the way in which it is constituted. There are two main types, "raw" glaze and "fritted" glaze. Raw glazes are prepared from materials which although having a small particle size are not soluble in water to any great extent. Frit batches contain some materials which are water soluble and these are consequently converted at an early stage of the glaze making process into a non-soluble frit or glass by heating with other suitable raw materials.

iii *Ceramic substrate.* A different method of arranging the class to which glaze belongs uses as an adjective the type of ware, body or substrate to which the glaze is applied. The definitions, therefore, become, for example, china glaze or earthenware glaze. To this can be added a subsidiary definition qualifying the glaze as being either "once-fire" or "twice-fire". "Once-fire" glaze is applied to a clay article and both body and glaze mature together in the same firing. "Twice-fire" glaze is a misnomer in that the glaze is applied to the fired body in the biscuit state and the chosen, second, firing cycle matures the glaze. It is thus the body that is fired twice. Variants of this system are soft fired biscuit hard fired glaze or hard fired biscuit soft fired glaze.

iv *Characteristic feature.* This distinguishing method names the most obvious characteristic of the glaze after firing and so refers to say a transparent glaze,an opaque glaze or a reactive glaze. Colour can be included if required.

v *Firing temperature.* In broad terms a glaze might be said to be a low firing type or a stoneware glaze, but it is more usual to designate the glost temperature referring either to the firing temperature in degrees celsius, to a cone number or to Bullers ring number. The following types can be included in this classification:

a Low melting glazes maturing in the range from 600–900°C for special applications, usually in the field of electro-ceramics.

b Majolica glazes maturing in the range of approximately 900–1050°C. This type of glaze can be leadless or lead-containing, and they usually contain B_2O_3. The term "majolica" was once used to describe a type of red ware glazed with an opaque tin glaze. Today such a term indicates only that the glaze is one with a low maturing temperature.

c Earthenware glazes covering the range 1000–1150°C. Again these glazes can be lead-free or can contain lead, but for most purposes earthenware glazes are usually low solubility compositions.

d Fireclay and sanitary glazes maturing in the range 1180–1250°C. These glazes are invariably of the leadless raw, unfritted type.

e Porcelain glazes maturing at 1300°C and above.

In general the glaze technologist combines those aspects of each method of classification which suits the circumstances at the time he is preparing the glaze to conform to the specification.

Chapter 3
Raw Materials

3.1. CHOICE OF MATERIAL

Raw materials for glaze and frit manufacture are selected from processed and beneficiated minerals and from bulk industrial grade chemicals.[1-3] The relatively low cost of glaze is based on the fact that its component raw materials are plentiful, are easy to obtain and are economically priced. Factors considered when assessing the suitability of a particular raw material for making vitreous compositions on a large scale are:

1. Chemical composition and its consistency over a long time.
2. Cost.
3. Mineral impurities.
4. Grain size.
5. Behaviour in storage.
6. Behaviour during mixing.
7. Behaviour when melting in furnace or kiln.
8. Location of source.
9. Availability.
10. Behaviour in suspension in water and on storage in this condition.
11. Effect of use on environment.

These factors will vary in importance according to the type of glaze being made. While no single factor can be ignored, the purity of the raw material and its cost will have a major influence on selection. Raw materials for glaze or frit are rarely represented by an unequivocal chemical formula and they usually contain some impurities. The utmost importance should be attached to this aspect of the quality of the material to prevent the introduction of undesirable impurities. All materials contain some iron oxide which may be at a tolerable level, but some possible sources of important glaze and frit components also contain iron chromite, zircon or rutile in amounts too significant to ignore. However, it is the major element oxide composition of the various glaze-making raw materials which are of prime interest.

When examining the different ways of introducing, say, alumina into a silica-based frit the cost of all the individual raw materials which are available must be considered. Unless there is an advantage to be gained, such as purity of tone or the need to have freedom from added alkali, then the use of calcined alumina and silica will be found to be uneconomic as source material compared with the use of a clay, felspar or a felspathoid. The choice made of these alumino-silicate minerals might lead to the elimination for the need to add separately some alkali and savings would accrue from the avoidance of a separate fritting operation.

Uniformity of grain size means in the early stages of manufacture, controllable mixing and frit melting[4] and the later grinding cycles can be standardised. Silica represents one of the major refractory raw materials in most frit compositions and its grain size can strongly affect the energy required for melting and complete solution. Finer grains will melt with less energy input, but particles which are too fine can create many small diameter bubbles which require additional heat to remove. In addition there is the important aspect that fine silicas are a health hazard. Any coarse grains which take longer to melt form areas of high viscosity in the frit, wherein bubbles of the reaction gases remain trapped or adhering tenaciously to the crystalline silica. The compromise is for frit-making raw materials to be substantially minus 0.5 mm (30 mesh BSS) grain size.

Knowledge of the behaviour of raw materials in storage is needed. Multi-compartmented silos will be chosen to ensure regular turnover for stocks of non-inert materials. Hygroscopic materials must be kept in warm and dry conditions. Friable materials should not be subjected to unnecessary handling.

The location of the source of a desirable raw material has a direct effect on the purchase price of the commodity, but the technical benefits of using, say, a fritting sand from a distant source (for example Belgium) rather than an indigenous sand can compensate for the cost differential. Where the material sources are overseas, then local political considerations might be involved when making our choice, but if the supply is a traditional one and the reserves are adequate, a balance might be made of the risks involved in establishing, on an uncertain base, a complex glazing process.

Raw materials can be added to the mill for further reduction in particle size or merely incorporated efficiently in the glaze mixture, but they must not affect the rheological properties of the slip to an undesirable extent over a fixed period of time.

There exists in modern industry a marked awareness of hazard, arising from any part of the many standard processes and consequently the selection of some (preferred) raw materials has been affected. Fluorine is a volatile element which is of value in reducing the viscosity of glassy melts, but its use has been greatly restricted by legislation relating to effluent control. Evolutions from frit plants and from glaze maturing in glost kilns has been reduced to almost zero through the substitution of china stone by beneficiated felspars. (China stone contains small amounts of fluorine minerals.)

Where high-lead frits are melted on a small scale, the use of granular lead glasses produced in bulk by specialist manufacturers is preferred to the use of other lead chemicals in powder form. It is accepted that the choice of finely ground materials can reduce the amount of energy needed for melting and because this reduces furnace temperatures and the volume of hot gases, there is a reduction in the amount of particulate matter entrained in the waste gas. For safety reasons arsenic, once used as an opacifier, is now no longer used. Salt glazing is practised in only limited areas of the world because of the problems associated with the gaseous effluent from the kilns.

Most products are received in the factory in packaged form—usually paper sacks—and the safe disposal of these should be considered under environmental impact.

In the sections which follow, where the practical raw materials are described, the stability in storage is to be accepted as one of the reasons for a materials's consideration as a usable component. Where this stability is not the norm, reference will be made to the storage difficulties.

3.2 AVAILABLE RAW MATERIALS

3.2.1. Silicon

Silicon dioxide (SiO_2) can form a glass on its own, but it makes an impractical glaze. With other constituents it is used in varying amounts (occasionally only a few per cent) in frits and glazes. The majority of ceramic coatings in current use have contents within the range 45% SiO_2–80% SiO_2.

Silicon is the second most abundant element in the continental crust at 28.2% and its oxide, silica, is hugely abundant in nature, occurring in seven distinct mineral forms.[5] Quartz is the commonest form and the one most widely exploited. The other varieties are tridymite and cristobalite, which are widely distributed in volcanic glass and opal, coesite, stishovite and lechatelierite. Silica occurs in combination with other oxides in minerals such as the felspathoids and other silicates.

Large deposits of quartz occur in quartzite,[6,7] a rock in which the constituent grains have recrystallised in an interlocked texture. Sandstone is a recemented product of weathered quartz. Ganister is a useful economical source of a good quality silica. Silica sand[8] is an assemblage of individual grains. Such sands can be formed by the weathering of sandstone or quartzite or prepared by mineral flotation processes. Cryptocrystalline silica is found in nature as flint,[9] a compact black silica, and also as chalcedony, jasper and opal.

Commercial deposits of high-grade silica for glaze and frit are relatively scarce and therefore the source of silica which is used[10,11] has usually been subjected to a degree of beneficiation or upgrading by a physical or chemical treatment. Mineral grains other than quartz are often present, but not all are detrimental. Felspathic minerals are themselves used as a source of silica[12] in vitreous compositions.

When preparing frit and glaze recipes it is usual to introduce silica where possible as a component of other ingredients. Felspathoidal minerals, when used as a primary source of silica in frits and glazes, also contribute some alumina and some alkali metal oxides in an economical form.

Silica is the major glass-former and occasionally it is the only network-former used in frits and glaze. The higher the proportion of silica in the glaze, then the higher will be the glost temperature, but the benefits from this apparent disadvantage is the great resistance such a glaze has to mechanical damage and chemical attack. In low-temperature glazes the ratio of silica to fluxes is about 2:1 (mol.). For the highest temperature glazes the ratio rises to 10:1 (mol.). An increase in silica reduces the coefficient of thermal expansion of the glaze and the fluidity of the glaze at its maturing point. Silicas from different mineral origins behave differently in frits and glazes because associated impurities act as fluxes and colorants.

Silica "Sand" (SiO_2)

Deposits of silicon dioxide, silica, are widespread and their sites are well known,[13,14] although because of impurities the sands are rarely used in an untreated condition. The mineralogical impurities are of two kinds, inert refractory minerals and minerals which produce unwanted colours in the glaze, notably those containing iron and chromium. Sands are upgraded in quality by mechanical tabling (Wilfley table)[15], by acid washing[13,14] and by froth flotation[16].

Iron oxide contamination can be removed from the surface of sand grains by washing with sulphuric acid and occasionally by treatment with hydrochloric acid or hydrofluoric acid.

Some silica sands contain appreciable amounts of felspathic minerals and allowance for the accompanying metal oxides (alumina and alkali) accordingly must be made in frit or glaze recipe.

Quartz (SiO_2)

Quartz occurs in veins or dykes in igneous rocks as a result of crystallisation from a liquid magma. It is consequently more difficult to win and prepare for glaze use than are silica sands. Some quartz deposits are extremly pure and are valuable for certain special high-purity glazes such as those which mature in reducing conditions.

Flint (SiO_2)

This dark coloured, though pure, crystalline form of silica is associated with chalk deposits and weathered flint pebbles are often found on the seashore. Flints for the ceramic industry are largely a byproduct of the cement industry.

Calcination of the flints at 900°C removes entrapped organic matter and water from the flint, thus opening up its dense structure and permitting easier grinding.

On health grounds this source of silica should not be used as a *dry* component of frit batches nor as a mill addition in the grinding of glaze.

"Silicates"

The mineral silicates discussed in other sections are all contributory sources of silica.

There is a range of alkali silicates (sodium and potassium) available as a prepared glass having widely different ratios of alkali to silica. All are water soluble to a degree and consequently have only a limited value in glaze technology.

Cullet is a soda–lime–silica glass available for recycling. Its source should be known and maintained to ensure constancy of composition.

3.2.2 Aluminium

Alumina cannot form a glass alone, but as an intermediate oxide it can take part in the network. In glazes, free from silica, it is usual to find some aluminium in the composition and in the simplest state it is reported[17] that with rapid cooling techniques (which affects boundaries of glass-forming areas) glasses of calcium aluminate have been formed.

Aluminium is the third most abundant element (by weight)[18] in the continental crust at 8.2% and, by inference, minerals containing aluminium should be plentiful. Felspars are a major component of the earth's crust and if other aluminous rocks are included the total of possible minerals in the crust which might be considered for ceramic use reaches 70% of the total volume. Frit and glass recipes can, therefore, include felspars, nepheline syenites, cornish stone, granites, pegmatites, aplite, clays, bauxites, diaspore, trachytes, phonolites, talcs, basalts, amblygonite, lepidolite and petalite. Commercial deposits of these minerals are exploited, but whether as a component of frit or glaze the source of alumina is usually either a clay or a felspar. In special compositions alumina trihydrate ($Al(OH)_3$) or calcined alumina (Al_2O_3) is chosen because of the high level of purity.[19]

The presence of aluminium in a glaze or frit introduced as either a pure chemical or a mineral component affects the chemical durability, mechanical strength, thermal expansion,[20] viscosity and surface tension[21] of the molten glaze, crystallisation and the behaviour of the glaze surface in the glost fire. In leadless, borosilicate glazes a small amount of aluminium exerts a considerable effect on thermal expansion in comparison with its effect on other types of glaze. For maximum resistance to attack by water, soda–lime–silicate glasses should have an aluminium content of one-eighth of the alkali content, but most glazes have more. Experience has shown that for earthenware glazes the molecular ratio of Al_2O_3:SiO_2 should be in the range 1:6 and 1:10.

In glazes where some control is exercised over the growth of a crystal phase the optimum ratio Al_2O_3:SiO_2 is 1:5. Viscous glazes develop finely textured matt surfaces because of the limitations on the development of coarse crystals. Increasing the aluminium content of a glaze raises the maturing temperature. Conversely, addition of aluminium oxide can eliminate unwanted crystallisation or devitrification.

The presence of aluminium is desirable in glazes associated with pink decoration based on the manganese–alumina and chrome–alumina pigment systems.

For many types of glaze the aluminium content covers a wide range because it is there for reasons other than promotion of resistance to attack by water. The structure of fritted lead silicates is strengthened by the addition of 2% (wt) alumina, thereby raising the resistance of the frit to attack by organic and body acids.

How these benefits derive can be deduced from the simplified two-dimensional silicate structure.[22,23] The aluminium ion can be four- or six-co-ordinated and in a silicate glass the aluminium Al^{3+} ion replaces the Si^{4+} ion in the structure forming AlO_4 tetrahedra. To maintain a balance other positively charged atoms, say sodium ions, are introduced. All the aluminium ions can be accompanied by an equal number of sodium ions in adjacent voids in the structure. This arrangement is found in many alumino-silicates. It is believed that aluminium oxide takes part in the random glass network in a similar fashion.

Aluminium Hydroxide, "Alumina Trihydrate" ($Al(OH)_3$).

Gibbsite is a pure, white, alumina trihydrate extracted from bauxite. When finely ground it has good suspension properties in the glaze slip and it can replace a proportion of clay in the recipe of any glaze which is prone to cracking on drying (i.e. shrinkage).

Alumina (Al_2O_3)

Aluminium oxide exists in nature as corundum. Aluminium oxide, alumina, is a product of a highly developed technology being prepared from the transformation at high temperature of precipitated aluminium trihydrate. As an ingredient in frit recipes it is sometimes preferred to alumina hydrate because it is more readily dissolved during the glass-forming reactions.[26]

"Clay"

Clay minerals mostly have a layered structure based on sheets of silica tetrahedra. They have a finely divided platey crystalline form, with a particle size of less than 2 μm.

There are widespread deposits of clay,[24,25] though many are unsuitable for use in glaze. Where it exists in large deposits the particles of clay are usually smaller than adjacent minerals and concentration can be carried out by washing the several materials from their deposits into a slurry for beneficiation. China clay or kaolin is a major mineral constituent in ceramic bodies and glazes. It has two principal functions; firstly in the frit as a readily assimilated source of aluminium and silicon, and secondly as the rheology control agent in the glaze slip. Ball clays are the preferred suspending agents for glaze slips applied to clayware. For glazes where fired colour is not important, lower grade ball clays, impure "red" clays or fireclays can be used.

"Felspathic Minerals"

Under this name is grouped two of the commonest rock-forming silicates of prime value to the glaze technologist, felspars and felspathoids. The felspars are relatively simple alumino-silicates, orthoclase ($KAlSi_3O_8$) albite ($NaAlSi_3O_8$) and anorthite ($CaAl_2Si_2O_8$). The felspathoids crystallise from magmas deficient in silica and the commonest is nepheline ($NaAlSiO_4$). They are abundant, though they do not exist as workable deposits on their own. Nevertheless, felspathic minerals are readily separable from other minerals and they are in great demand as an inexpensive source of aluminium and alkalis.[27–29]

Most of the world's pure felspar is extracted from pegmatites by froth flotation processes. Nepheline syenite, a coarser grained rock, is beneficiated by dry, high intensity magnetic processes.

Anorthosite[30]

The plagioclase felspars, solid solutions of albite ($NaAlSi_3O_8$) and anorthite ($CaAl_2Si_2O_8$), are the commonest mineral series. The rock anorthosite consists mainly of these minerals in proportions depending upon the site of the deposit. The calcium content can therefore vary between wide limits and commercial grades contain between 8–23% CaO. It is used as a source of aluminium oxide with valuable contributions to the alkaline earth components in frit or glaze.

"Pegmatite"

Pegmatites are rocks of granite composition containing less alkali than does an upgraded felspar. They have a higher iron oxide content and are consequently used only in certain types of glaze. Granite itself has been used similarly in some glazes.

"Stone, China Stone or Cornish Stone"

A partially kaolinised granite, cornish stone, is used in limited quantities in frits and glazes. It is of variable composition and its quality is graded according to its fusibility. The alkali content ranges from about 5% to about 8%. The most fusible, containing perhaps 8% (wt) alkali (total), also contains a proportion of fluorspar (CaF_2). Whilst acting as a powerful flux, fluorspar unfortunately can subsequently mar the glaze surface with pinholes due to its variable degree of decomposition during the glost cycle.

Prepared mixtures[31] of minerals with closely controlled properties and free from fluorine are available as substitutes for natural stone in both body and glaze recipes.

3.2.3 The Alkali Metals

Compounds of the alkali metals, lithium, sodium and potassium, are widely distributed in nature as water-insoluble compounds in rock-forming minerals and are widely used in one form or another in industry. Other alkali metals are not so abundant nor so frequently found in use, but some felspars contain appreciable amounts of rubidium. Although simple water-soluble alkali salts occur in nature, most of the raw material of industry is chemically prepared. Being water-soluble, all the essential alkaline compounds needed for glaze must first be melted into glassy frits unless the required amount can be obtained from introducing felspars or felspathoids.

The glaze technologist will always consider carefully which is the best route to introduce any normally soluble component. Two glazes with identical analyses, ground to the same particle size under identical mill conditions but with different mill recipes, could still have different slip properties. Alkali introduced by way of a felspathic mineral is less water-soluble than if it was introduced as a component of the fritted part of the glaze. Glasses are more readily attacked by water than are many rock minerals and the alkali leached from the glassy frit acts as an electrolyte, so influencing the rheology of the glaze slip.

The prime function of the alkali metals is as fluxes. In general they lower the melting points of frits and glazes and, by increasing the fluidity at the glost temperature, enhance the development of gloss. The refractive indices of some frits can be increased by their inclusion. The thermal expansion of vitreous compositions is markedly affected by the inclusion of alkalis and their balance is largely used to control the crazing resistance of glaze. Changes in the electrical properties of silicate, borosilicate and borate glasses brought about by partial replacement of one alkali metal for another are so marked, however, that the effect merits the phrase "the mixed alkali effect".[32,33] These changes are accompanied by beneficial changes in other physical and chemical properties and can be of great value to a glaze chemist. An increase in a chosen alkali can increase chemical resistance in an apparently anomalous way and the vitrification range can be extended.

Lithium

The value of lithium in several important glaze functions encourages its wider use. Whilst its behaviour in glaze or frit is to be expected from its position in the series of alkali metals, some effects are anomolous. Thermal expansion can be reduced by an equimolar substitution of sodium or potassium by lithium and crystalline glazes with low expansions are easily formed.[34] It acts as a melt accelerator in frit making and rapidly reduces the melt viscosity at low levels of addition.[35] The melting cycle can therefore be reduced either in time or in temperature and it means that glazes can be designed to melt at lower temperatures without risk of crazing because the desired properties can be obtained[36] using a lower alkali content (w/w). The maturing time of lithium-containing glazes is shorter than for similar glazes containing the other alkalis.

The strong fluxing action allows the inclusion of greater amounts of alumina and silica to give high strength glazes with excellent durability. When partly replacing lead oxide there is less tendency for lead to vaporise during the glost fire.

Lithium forms contracted glasses, that is glasses which have densities higher than might be expected from calculation,[37] and this effect has found use in increasing the surface hardness of glazes. The replacement of PbO, Na_2O and K_2O by Li_2O increases the hardness of tableware glaze by 20%.[38] Whereas sodium and potassium tend to reduce surface tension, lithium behaves like the alkaline earths and increases it.

Supplies of suitable raw material for frit and glaze making are adequate. The particular lithium compound chosen[39] will depend on many factors. While it is more usual to add one of the four common lithium minerals, lithium carbonate would be chosen for frits with low alumina contents.

Lithium carbonate (Li_2CO_3). This is the purest source of lithium in glass and ceramics. Being only slightly soluble in water, lithium carbonate can be used in raw (alkaline) glazes.[40] Such glazes are good bases for some rich colours. Lithium carbonate is not hygroscopic and is readily stored under normal conditions without caking.

Lepidolite ($K(AlLi_2)Si_4O_{10}(OH)_2$). Lepidolite is a phyllosilicate often called lithium mica. It only occurs as a purple mass of small flakes. On average it contains 1.3–2.4% of lithium oxide. The specific gravity is 2.8–2.9. There is often present appreciable amounts of the other alkalis, e.g. rubidium.

Petalite ($LiAlSi_4O_{10}$). Petalite is a lithium alumino-silicate containing 3.5–4% lithium oxide by analysis. Its specific gravity is 2.42. Petalite is widely used and up to 35% can be added as a mill addition to a wide range of glaze compositions. It is usually greyish-white, but occasionally pink grades are found.

Amblygonite ($LiAlPO_4F$). This is a pegmatite mineral containing 8–9% of lithium oxide. Its natural colour in the massy rock form is white to grey and its specific gravity is 3.0–3.1.

Spodumene ($LiAlSi_2O_6$). This second lithium alumino-silicate (a pyroxene) exhibits a colour which varies from white in low iron grades to green in iron-rich crystals. Normal supplies for ceramic use average about 2.4–7% Li_2O in the rock. Its specific gravity is normally 3.1–3.2, but this changes markedly at about 1000°C to about 2.4 as a consequence of a 30% volume increase. This transformation to beta-spodumene is irreversible.

Lithium fluoride (LiF). An obvious double fluxing agent, lithium fluoride can be used to advantage in frit compositions. Being only slightly soluble (0.27% at 18°C) it can be used as a mill addition. It lowers the melting point and increases the firing range.

Lithium metaphosphate ($LiPO_3$) and lithium orthophosphate (Li_3PO_4). These have only a limited solubility in water and can be used in raw glazes.

Potassium

The properties of frits and glazes containing potassium resemble those of sodium-based coatings, but in detail are sufficiently different to merit care when considering substitution of one family of raw materials with the other. In general, the inclusion of potassium at the expense of sodium decreases the mobility of the glaze during firing and reduces the coefficient of thermal expansion. Crystal growth is not so active in potassium-based glazes. The resistance of frit to attack by either water or chemicals can be improved by the balanced substitution of sodium by potassium. Normally the source of the potassium needed in frits can be obtained from felspathic raw materials. The tints of some coloured glazes can be modified by its inclusion in the frit.

Raw materials for a source of potassium in frits and glazes are widely available.

Potassium bicarbonate ($KHCO_3$). The use of potassium bicarbonate is recommended[41] because of its stability in storage and its free flowing nature when fresh. It is not hygroscopic, but it nevertheless has a tendency to cake during prolonged storage.

Potassium carbonate (K_2CO_3). Potassium carbonate or potash is available either in the calcined form or hydrated with 15% H_2O. This latter is known as pearl ash

($K_2CO_3.1\frac{1}{2}$ H_2O). The calcined form is hygroscopic and storage of it should therefore be carefully monitored. Pearl ash is the more stable of the two compounds during storage.

Potassium nitrate (KNO_3). Potash nitre or saltpetre is available as a relatively pure chemical. Although not used as a primary source of potassium, it plays a valuable role as an oxidising agent and as an active flux (m.p. 334°C) in the early stages of frit melting.

Sodium

Sodium is perhaps the main alkali metal in glaze, although it is seldom used alone.

Tensile strength and elasticity are decreased as the content of soda increases, although to a lesser degree than with the other alkalis. Small increases of sodium oxide in a glaze will increase the coefficient of thermal expansion and reduce the softening point with rapid and noticeable consequences to the glaze/body fit. The addition of sodium decreases the durability of frit and glaze. Suitable raw materials are available in large quantities.

Sodium carbonate (Na_2CO_3). The ashes of littoral plants contain sodium carbonate as the major ingredient and barilla and kelp ashes were the source of soda until the introduction of the Leblanc process in the early nineteenth century. Because of its alkali content, wood ash is often used by studio potters as a basis for artistic glaze effects.

Soda ash or sodium carbonate can be obtained either from the vast deposits of natural brines such as trona ($Na_2CO_3.NaHCO_3.2H_2O$) or natron ($Na_2CO_3.10H_2O$) or as the product of the Solvay process in which sodium chloride, limestone and ammonia are reacted. The choice between the two sources is primarily one of availability. The Solvay process still accounts for about 90% of the world's production, but environmental restrictions in some countries are forcing a greater utilisation of the brine extracts. Supply is adequate and purity[42] is excellent from either source. The level of iron oxide impurity is negligible. Dense soda ash (more widely used for frit making than is light ash) is granular and normally dust-free, but its friability during rough mechanical handling can cause dusting. Prolonged storage in silo or sack leads to caking.

Hydrated sodium carbonate ($Na_2CO_3.10H_2O$) is rarely used in frit making. Much steam is evolved during the initial stages of melting which adversely affects the temperature distribution in the batch.[43]

Sodium chloride ($NaCl$). As "common salt", sodium chloride is abundant, cheap and has a consistent level of purity, but it has only a limited use in glaze technology and is not used as a source of alkali in the making of frits.

The vapour glazing of heavy clay products was once widely practised, but legislation controlling the permitted levels of gaseous effluents has curtailed the application of this technology.

Sodium hydroxide (NaOH). Use of caustic soda as a 50% solution has been suggested[44] as a source of sodium and as a means of consolidating briquetted batch prior to fritting. No known example of its use in frit manufacture exists.

Sodium silicofluoride (Na_2SiF_6). This reaction product of soda ash and hydrofluosilicic acid is a very active flux in frit manufacture, being used as a substitute for sodium fluoride and where it is necessary to add fluorine without the addition of aluminium or calcium. The enhanced reactivity comes from the evolution of fluorine early in the fritting process. It is of particular value in low-temperature glazes opacified with titania. It is not hygroscopic and only sparingly soluble in water. Decomposition starts at about 325°C and above red heat the decomposition is appreciable. *Caution*: sodium silicofluoride is a scheduled poison.

Sodium nitrate ($NaNO_3$). Sodium nitrate or Chile saltpetre has no place as a major source of sodium in frit or glaze. Both natural nitre and processed grades of sodium nitrate are available, but its high reactivity in contact with refractories (when molten above 308°C) prevents its use except in small quantities as an aid in the early stages of frit melting. It is a good oxidising agent, releasing oxygen slowly at temperatures up to 520°C. Care should be taken in storage to avoid "caking".

Sodium phosphate. One of a family of compounds normally used in water treatment, but sodium hexametaphosphate is used to control the viscosity of fluid glaze slips. It is more often used to add phosphate ions to a glaze frit rather than as a source of soda. The phosphate ion is an important colour control agent during the melting of titania opacified frits.

Sodium sulphate (Na_2SO_4). Sodium sulphate is not much used in glaze manufacture. It is found in natural brines as well as being a byproduct from some industrial chemical processes.

3.2.4 Alkaline Earth Metals

Barium

The use of barium in glass compositions was studied in the nineteenth century[45] and it was accepted that barium carbonate made the batch more fusible and the glass denser. The principal advantage of its inclusion at the expense of other alkaline earths in a frit or glaze is the increased "brightness" of the glost surface, deriving mainly from the development of higher refractive indices and partly from the additional fluxing power of barium components. In this, as well as in the small way it affects thermal expansion, its behaviour resembles that of lead; consequently it is often used as a substitute for lead. Barium does not have so marked an effect on specific gravity as does lead. For similar firing temperatures, the properties of barium frits (durability, elasticity, density or refractive index) fall between those of lime frit and lead frit. Because of the beneficial effect it has on elasticity and mechanical strength, compared with the other alkaline earths, barium is often found in glazes on bricks. It has a valuable role to play in the development of finely crystallised matt glazes.

The inclusion of barium in glazes and frits does, however, have some disadvantages. Care must be taken when using barium compounds because of their toxicity. Heavy corrosion of frit furnace refractories occurs when batches having a high barium content are melted.

Barium carbonate ($BaCO_3$). As the mineral witherite, barium carbonate is found in nature, but its present rarity precludes its wider utilisation. The most widely used frit ingredient is barium carbonate prepared from the mineral barytes. Two methods are used, one by precipitation with sodium carbonate and one by precipitation with carbon dioxide. When it is necessary to have minimal alkali content then barium carbonate from the second process should be chosen. Barium carbonate material is poisonous.

Barium carbonate is an active flux at high fritting temperatures, although when used in excess it can lead to undesirable crystallisation.

Barium chloride ($BaCl_2$). Barium chloride (poisonous) is a water-soluble compound sometimes used in place of calcium chloride as a rheological control agent. It is not used as a principal ingredient of either frit or glaze.

Barium sulphate ($BaSO_4$). Barytes is the well-known mineral form of barium sulphate. In nature it is abundant; consequently it is an economic source of barium in frits. Its use as a non-hazardous souce of barium as a mill addition needs the exercise of caution because of the excessive amounts of sulphur oxides which will be evolved during the glost fire.

Barium nitrate $Ba(NO_3)_2$. The main use for barium nitrate is as a frit ingredient. It is an alternative to the usual alkali nitrates because in the furnace it does not decompose so readily, melting at 592°C. Being moderately soluble in water, care in handling is required because of its toxicity.

Calcium

Calcium compounds of widely differing compositions are used either as constituents of the frit batch or as mill additions in glaze formulations. Benefits are derived from their presence in the early stages of heating when calcium compounds readily react with other common glaze components. Later, at higher temperatures, the fluxing ability of calcium oxide is good and its effect in reducing the viscosity is very marked. However, low amounts are desirable in glazes which are required to have low viscosity at enamel firing temperatures, and therefore assist in ensuring interaction between glaze and enamel colours.

Calcium oxide increases the mechanical hardness and tensile strength of the glaze and improves its durability to acids and to water. In a glaze with a balanced composition an increase in calcium oxide can reduce the coefficient of thermal expansion. An excess of calcium oxide, however, increases the tendency of the glaze to devitrify. If the crystallisation of anorthite ($CaAl_2Si_2O_8$) and wollastonite ($CaSiO_3$) is deliberately encouraged, then a range of glazes with varying degrees of mattness can be produced. Unlike the other alkaline earths, calcium oxide does not greatly affect the refractive index (and so the "brilliancy") of glaze.

The calcium compounds in general use are abundant, consistent in composition and economic.

Calcium carbonate ($CaCO_3$). Several mineral forms exist and unfortunately their names are used indiscriminately to describe the different types. Limestone, a sedimentary rock, predominantly calcite, is the main source of raw material because it is readily available in a pure form.[46] Marble and chalk are not so widely used. Precipitated calcium carbonate, a byproduct of the water industry, is uniform in composition and has a small grain size. Whiting has a specific meaning of a finely ground mineral and as a name is often used to describe grades of pure powdered limestone. For the ground mineral a wide range of different particle sizes is used from a coarse 6 mm grain size to a 75 μm (200 mesh) grain size. The main impurities in limestone have been introduced during its deposition (it constitutes about 30% of all sedimentary rocks) and quality will be assessed according to the presence of small clastic quartz and silicate mineral grains.

The mining of limestone is an ideal route to avoid contamination from overburden during quarrying. Aragonite, coral and seashells are sources of calcium carbonate used, where available, for frit manufacture.

Burnt lime (CaO) or burnt dolomitic lime ($CaO.MgO$) are not used, although their reactivity with other batch materials during melting has long been recognised.[47] Hydrated or slaked lime, $Ca(OH)_2$, has a limited though specific value as a glaze component (see Section 11.4.2).

Calcium chloride ($CaCl_2$). Although not used as a major ingredient, calcium chloride plays an important part as a flocculant in the control of the fluid properties of the glaze slip. The viscosity of the slip will be increased by the addition of this electrolyte.

Calcium fluoride (CaF_2). Calcium fluoride occurs as the mineral fluorspar or fluorite. Three qualities are available graded according to the calcium fluoride content, and there are many sources with different particle size.

Calcium fluoride is a very active flux, particularly in the early stages of melting. It is employed only sparingly as a source of calcium, but is used to increase fluidity in the molten glaze. Fluorine is volatile and consequently gives rise to bubbling during the glost fire. Despite this disadvantage, the presence of fluorine ions in a glaze is beneficial to the development of some colours.

Calcium phosphate ($Ca_3(PO_4)_2$). Calcium phosphate has only a small usage in glaze compositions. It is the principal component of bone ash and occurs in the mineral apatite. Bone ash also contains some free lime ($Ca(OH)_2$) for which an allowance should be made in calculating recipes.

Anorthosite. See Section 3.2.2.

Calcium silicate ($CaSiO_3$). Calcium silicate occurs in a pure form as the natural mineral wollastonite. There is a series of different calcium silicates, many of which are formed in slags and cements in an impure form and are unusable for glaze.

Despite its small solubility in water, wollastonite is often the preferred source of calcium ions in matt glazes, producing a smoother texture than could be obtained by other calcium compounds. The development of an artificial wollastonite with a precise crystalline form for bodies gave to the glaze technologist an alternative and readily controllable source of a useful raw material, though for this purpose the crystal habit is not important. It is not usually used as a frit ingredient because of the availability of more economically priced sources of CaO and SiO_2.

Calcium sulphate ($CaSO_4$). Calcium sulphate can be found in nature as the mineral gypsum ($CaSO_4.2H_2O$) or as anhydrite ($CaSO_4$). There is only a limited use for either material as a source of calcium oxide in glaze because of problems related to gas release, but there follows from the use of either, two important subsidiary functions in glaze manufacture. Firstly, in frit melting an addition of say 1% to the batch in the furnace will increase the rate at which reaction gas bubbles are released from the melt. Secondly, calcium sulphate, often as "potters plaster" ($CaSO_4.\frac{1}{2}\ H_2O$), is sometimes used as an electrolyte in the control of the flocculation state of glaze slips (see Section 11.4.2).

Strontium

Prior to 1942, strontium compounds were not widely used, although their value in improving melting rates has long been known. The use of strontium in conjunction with barium and calcium gives benefits beyond what might be expected from the mixed base effect[48] when there is need to control crystallisation in glazes. In direct comparison with barium oxide alone, strontium oxide produces brighter and more fusible glazes.[49] Strontia is known to improve the chemical resistance of glazes when introduced at the expense of lead oxide and zinc oxide, but its primary role is to facilitate the development of low-temperature glazes.[50,51] Using strontium oxide as the flux, glazes with high gloss, firing at 1145°C (cone 01) can be made with good crazing resistance due in part to the highly developed glaze–body reaction layer. Strontium in glaze can eliminate surface textural defects characteristic of barium glazes fired in an atmosphere high in sulphur.

Care should be taken when combining calcium and strontium compounds in a glaze, because strongly crystallisable compositions can readily be made.[52]

Recent investigations have concentrated on the non-hazardous nature of strontium-based glazes. Satisfactory fritted and raw glazes, free from lead oxide and zinc oxide, can be made[53,54] spanning a wide firing range from 800°C to 1300°C.

The commercial aspects of the use of strontium compounds are related to those of calcium and barium because these elements have similar properties and they are often substituted for strontium. The common natural forms of the three elements, sulphate and carbonate, show similar properties. The most important group is the sulphate represented by anhydrite ($CaSO_4$), celestite ($SrSO_4$) and barytes ($BaSO_4$). All three minerals are colourless to white when pure.

Strontium carbonate ($SrCO_3$). Strontianite, the natural strontium carbonate, is available, but it is the prepared strontium carbonate that is the preferred ingredient for frit or glaze. This has the advantage over barium cabonate of being non-poisonous. It can be used either as a mill addition or in raw glazes despite its slight solubility in some grades of water.

Strontium sulphate ($SrSO_4$). Celestite, natural strontium sulphate, is the principal strontium mineral in current commercial use. It is the source ore for the preparation of strontium salts. Small additions to frit batch are an aid to early glass-forming reactions in the kiln.

Magnesium

Magnesium compounds are present to some extent in limestones, and consequently magnesium oxide can be found in most glazes when analysed. It rarely deposes calcium oxide as a major component except in a few special glazes. Like calcium it has an active fluxing action at high temperatures and it also reduces the melt viscosity, being most effective in the region of 1200°C. When adjustment is needed to the shape of the thermal expansion curve, magnesium added to the glaze can lower the annealing point. It improves adherence between glaze and body. The late fusion induced in glazes by high quantities of magnesia is used to produce mottled surface effects.

The development of colour in some glazes is affected by the inclusion of magnesium. It is particularly effective in producing cobalt purple and nickel greens. When magnesia glazes are fired in reducing atmospheres there is a marked development of a grey colour.

Source materials for magnesium oxide are abundant.[55]

Magnesium carbonate ($MgCO_3$). Natural magnesite, $MgCO_3$, is not widely available, but magnesium carbonate is available as a precipitated grade of $MgCO_3$ and as one of several basic salts. They are stable in storage up to 50°C. All have a fine particle size, which makes thorough mixing prior to frit melting a difficult operation; consequently the source of magnesium in frits might be chosen from seawater magnesite, dolomite or talc. Magnesium carbonate decomposes at 350°C.

Precipitated grades of magnesium carbonate can be used as mill additions despite their fine grain size, slight solubility and low bulk density. These factors lead to high shrinkage as the glaze dries and this can result in cracking and crawling during the glost fire. An excessive degree of shrinkage can be utilised in decorative glaze texturing.

Calcined magnesite or seawater magnesite (MgO). Primarily produced for the refractory industry, calcined magnesite is readily available in different grades of quality and grain size. The granular grades are preferred for introducing magnesium to the frit. The solubility of magnesite in water is dependent upon the degree of calcination.

Magnesium sulphate ($MgSO_4.7H_2O$). This is another example of a material not used as a principal source of magnesium but as an electrolyte in the control of glaze slip properties. It is occasionally used as a stoneware glaze flux.

Dolomite ($CaCO_3.MgCO_3$). Dolomite is the double carbonate of calcium and magnesium, usually formed by the alteration of limestone by solutions containing magnesium ions.

The compositions of commercial dolomite vary about the theoretical figure for $CaMg(CO_3)_2$, although within any deposit the mineral composition is consistent. Very pure grades are available with high levels of magnesium and low levels of iron oxide. Deposits formed by precipitation of mixed carbonates of calcium and magnesium are not as common.

Talc ($Mg_3Si_4O_{10}(OH)_2$). The mineral talc is not widely used in glaze manufacture, but its value lies as an economic stable source of magnesium free from calcium. Lamellar grades should be chosen.

3.2.5 Titanium

Titanium occurs in the mineral forms rutile and ilmenite and these are the principal source materials in glaze manufacture. Sphene and leucoxene are minor sources of titanium in some ceramic processes. In igneous rocks titanium is mainly present as ilmenite. Titania is also widespread throughout many clays and sedimentary rocks. Pure, pigment grade, titanium dioxide is used in low-temperature vitreous coatings.[56]

The principal role of titanium in frits and glazes is to increase resistance to acid attack. Titania is an excellent opacifier of low-temperature glazes, but control of the degree of whiteness at temperatures above about 800°C is difficult and it scarcely competes with zircon in this application. A wide variety of crystalline effects (the "rutile break-up") is possible by adding mineral sources of titania to the pre-milled glaze slurry prior to firing. Titania is a potent nucleating agent in glass/ceramic glazes when incorporated in the frit. Glazes containing titanium are photosensitive. When present in sufficient quantity, titania increases the refractive index, but at the same time the development of colour would be markedly affected. Small amounts aid the improvement in crazing resistance, but above 2% addition there is little extra effect.[57]

Titania glazes "transmute" colour from the underlying body. Iron oxide in the body is absorbed into the crystals of titania as these form during the cooling cycle.

Ilmenite ($FeTiO_3$). This intensely black mineral is the source of pure white titanium dioxide pigment. It is found in large quantities in detrital deposits such as beach sands where it is recovered by dredging and beneficiation. In glaze it causes a surface break-up with variegated coloration. When added in the unground form to a glaze it produces speckled, textured glaze surfaces.

Rutile (TiO_2). Rutile, a reddish-brown to black mineral, occurs in many detrital deposits, particularly in beach sands. These are worked by dredging or scraping prior to treatment by heavy liquid separation[58] and flotation.[59] Iron oxide is an integral part of the crystal structure. Chromium and vanadium are occasionally present. Rutile has a poor pigmenting effect and its main value is in textured glazes. In conjunction with other oxide colorants, "break-up" and mottled effects are produced. Material with a wide range of particle size can be used.

Sphene ($CaTiSiO_5$). Sphene is a greenish-yellow to brown mineral found in association with ilmenite. It is little used in glaze as a mineral addition, but crystallisation of sphene is encouraged from glazes containing much calcium and titanium so as to produce an effective opacifying phase.

3.2.6 Zirconium

Zirconium does not enter into the common rock-forming minerals to any degree, but appears as a specific phase, usually zircon. It occurs in most igneous rocks and more abundantly in granites and nepheline syenites. Where these have weathered, large detrital deposits of zircon-rich sands are formed. Baddeleyite, a mineral form of zirconium oxide occurs naturally in a range of purities. Present supplies originate as a by-product of phosphate rock processing and its value as an opacifier has been known for some time.[60] Hafnium always occurs, camouflaged, in zirconium minerals.

Glazes containing zirconium range from full transparency to high opacity, from completely glossy to fully matt and are used on all types of ceramic substrate. It is as an efficient, economic opacifier that zirconium silicate is primarily used.[61] The degree of opacity which develops depends upon the relative proportions of other glaze components such as aluminium and zinc. X-ray examination of fired glazes shows zircon as the major opacifying phase when mill additions of either zircon or zirconia are made (see Section 7.8). When used in a frit, the opacifying phase is initially zirconia, which on heat treatment changes to zircon.

In its effect on the properties of a glaze, zirconium is analogous to aluminium. As the level of zirconium increases, even at the expense of silicon, the glaze becomes more refractory and has less flow at the glost temperature. Such glazes set very rapidly from the molten state. Increasing the amounts of zirconium compounds added to a glaze increases the density; at the same time the thermal expansion is lowered with a consequent improvement in the crazing resistance.[62] (The coefficient of thermal expansion of zircon is 4.5×10^{-6}°C^{-1}, which is low in comparison with other crystalline materials.) The mechanical strength of the glaze is improved with increasing levels of zirconium. With the addition of zirconium, frits and glazes with high refractive indices are readily prepared. Glazes containing large quantities of zirconium are very resistant to chemical attack and small amounts, insufficient to cause opacity, are frequently added to glazes to improve their durability.

It is an efficient opacifier in a correctly blanced glaze whether used as a mill addition or when melted in the frit component of the glaze. More zirconium silicate than tin oxide is needed in a glaze to produce an acceptable degree of opacity, but this is offset by an advantageous price structure. As an opacifier, zircon does not produce the pink discoloration often exhibited by "white" tin opacified glazes in the presence of an atmopshere containing volatile chromium compounds. Specking in zircon glazes is primarily a fault originating during fritting. Specks of carbon (say from badly set burners) and other contaminants have difficulty in being oxidised and eliminated from zircon glazes because the high intrinsic viscosity of such glazes allows little physical movement within the glaze layer even at very high glost temperatures.

Zircon ($ZrSiO_4$). Zircon is found in large volume deposits in the beach sands of India and Australia, where the deposits are the result of weathering of the parent rock and a natural concentration due to its high density of 4.6. The beach sands are obtained by dredging or mechanical scraping. A floating dredger picks up the sand and separates inboard quartz and rutile from the zircon. From a wide range of available particle sizes grades of zircon can be chosen to give the best performance as a frit component or as a mill addition.

Baddeleyite. Essentially zirconium oxide, baddeleyite contains iron oxide, titanium oxide and silica as impurities, but most grades can be beneficiated. The range of purity in available mineral sands is from 84% to 99% ZrO_2. It is less widely distributed than zircon and is usually extracted as a by-product from other mineral processes. The pure grades are upgraded by magnetic separation and by chemical treatment.

Zirconium oxide (ZrO_2). The prepared oxide of zirconium for incorporation in frits is available as a dense, free-flowing powder with high purity levels. The high reactivity of the chemically prepared oxide promotes rapid reaction in the frit furnace. There are several methods of producing zirconia from the mineral precursor. (Zirconia exists in three polymorphs (monoclinic, tetragonal and cubic), but these are not relevant when ZrO_2 is used in fritting.)

3.2.7. Cerium

Compounds of cerium do not dissolve easily in either frit or glaze and the principal use of cerium is as an opacifying phase. In low-melting-point glaze it is a high-intensity opacifier and a very effective alternative to tin oxide. The levels of opacification are helped by high alumina contents which further lower the solubility of ceria.

Combined with titania (2.5%), cerium oxide (2.5%) produces pleasant yellow colours[63] which are stable in oxidising and reducing conditions. Whilst this colour is a useful addition to the glaze palette, the presence of titania as an impurity in other ingredients is a disadvantage when using ceria as a whitening agent.

Cerium oxide (CeO_2). Cerium oxide is the only stable oxide and it is prepared from cerium hydrate. Its refractive index is 2.3. For use as an opacifier, only the purest grade, free from other rare earth oxides, is selected. It is only slowly soluble in either glaze or frit; consequently it is rarely melted in a frit but used mainly as a mill addition. It is an oxidising agent when it is utilised as a frit ingredient. As a mill additive it has a stabilising effect on yellow base stains using vanadium or antimony as chromophores.

3.2.8 Fluorine

Fluorine occurs in nature almost exclusively in a chemically combined form as a fluoride. The most important natural compound is fluorspar, and cryolite is a close second. The fluorine content of many rocks processed for other chemicals can be high and fluorine–silicon compounds are obtained as important by-products. It is present in some pegmatites (for example, "Cornish Stone") and in some clays.

Beryllium fluoride can form a glass,[64] and the structural similarity between BeF_2 and SiO_2 has been noted. There is a close similarity between the ionic radius of oxygen (1.40 Å) and fluorine (1.36Å), and many other glasses have been patented using fluoride systems rather than an oxide system. Fluorine is a potent agent (even in small quantities) for reducing the viscosity of frits. Used in larger quantity it causes bubbles in a glaze. However, it is sparingly used because of the ease with which it volatilises from melts, no matter which compound is used as its source. Excessive quantities of fluorine in the presence of high aluminium contents can produce opaque frits,[65] although such frits do not produce completely opaque glazes. Because of its fluxing ability the amount of added alkali can be reduced in a composition and therefore it can be said to lessen the tendency for a glaze to craze.

Finely dispersed bubbles of fluorine gas produce a degree of opacity, but this technique is not a viable alternative to other methods using solid phases for the opacification of glazes.

Fluorine is a very corrosive gas and is a very aggressive reactor with refractories and kiln furniture. In countries exercising control over pollution, fluorine is a listed effluent in all forms. All water-soluble fluorides are poisonous.

Calcium fluoride (CaF_2). Fluorspar is the most frequently used material as the source of fluorides in a frit or glaze with its addition being of particular value in low-temperature glazes. Rarely is it used as a primary source of lime, but it is used in small amounts to reduce the viscosity of frits without influencing any developments of opacity.

Aluminium fluoride (AlF_3). This white, often coarsely ground, material is used both as a flux and an opacifier in frits. It is stable in storage, slightly soluble in water and it sublimes above 1000°C.
Sodium fluoride (NaF). This poisonous though non-corrosive white powder is a widely employed flux, but its high alkali content often precludes its use alone as an opacifier. It is sparingly soluble in water (4 g in 100 ml) and stable in storage.
Sodium silicofluoride (Na_2SiF_6). This is a synthetic compound obtained from the neutralisation of hydrofluosilicic acid with sodium carbonate. It is a stable compound in storage if kept dry. As a flux and an opacifier it is widely used, often as a substitute for sodium fluoride.
Sodium aluminium fluoride (Na_3AlF_6). This double fluoride of sodium and aluminium is used as a flux (fusion point 950°C) and as an opacifier. In nature it occurs as the mineral cryolite which undergoes refinement by froth flotation and magnetting before being ground into different particle size grade. It is sparingly soluble in water and stable in dry storage conditions.

3.2.9 Boron

Boric oxide (B_2O_3) can form a glass on its own, but its dual use is as a flux and network former. The value of boron as a flux has long been recognised. Mention is made of borax as an ingredient in glass recipes of the seventeenth century in *L'Arte Vitraria*,[66] and some types of special glasses contain boron as an essential ingredient. However, until the end of the nineteenth century the supply of boron compounds was restricted. With its value in frit compositions proven, it has become exceptional for boron not to appear in recipes for low-temperature (i.e. $<1100°C$) glazes.

There are a great many boron minerals, but few which are commercially important. Tincal, borax, an impure sodium borate first traded from Tibet about 1563 for its value as a metallurgical flux, was obtained from lake mud deposits and desert deposits. Boron compounds in Europe came from thermal springs in Tuscany. There is now a wide range of boron chemicals available obtained mainly from California and Turkey. Ulexite, sodium calcium borate ($Na_2B_4O_7.Ca_2B_6O_{11}.16H_2O$) was first collected from borax marshes as a low-grade ore source of borax. Later a purer grade of ulexite associated with colemanite, calcium borate ($CaB_3O_4(OH)_3.H_2O$), was exploited until extensive deposits of borax and kernite (or rasorite) were found and became the principal sources of sodium borate. Dry lake deposits of sassolite are major sources of boric oxide. Other boron minerals are aschanite ($Mg_2B_2O_5.H_2O$), boracite ($Mg_7B_{16}O_{30}Cl_2$) and pandermite ($Ca_4B_{10}O_{19}.7H_2O$).

Boron has an important place in glaze technology and is the second most important network-former after silicon. It can participate in glass structures at two levels of coordination, in the form of BO_4 tetrahedra and BO_3 triangles. The change from BO_3 traingular form to BO_4 tetrahedra strengthens glass structure and brings it to a state resembling silica.[67]

The compounds of boron are rarely used raw in glazes, but rather as principal ingredients of many frits, their inclusion being almost wholly beneficial. Comparing boron sources for frit batch, fused (dehydrated) borax is found to be most advantageous, then boric acid and lastly, borax, that is in order of decreasing water content.[68] Practically all properties of a vitreous composition are changed by the addition of boron. Only if a threshold value of about 12% B_2O_3 is passed does the direction of the rate of change alter. This change has been called the boric oxide "anomaly", and it could indicate the point at which its coordination state changes.

As the temperature rises during the fritting cycle the presence of boron initiates glass formation at an early stage. The resistance of borosilicate compositions to chemical attack is good. High boron glazes mature rapidly as they are extremely fluid at glost temperatures, although the disadvantages to this property can be the ease with which phase separation can occur. Boron lowers surface tension, thereby increasing the speed at which smooth glaze surfaces develop. When added up to certain amounts it can reduce thermal expansion[69] and both mechanical strength and scratch resistance are improved after increasing the level of boron in a glaze. Boron glazes are good bases for glazes coloured by dissolved transition element oxides.

Salt glazing of stoneware and heavy clay goods is easier if borax is mixed with sodium chloride, because glazing then occurs at lower temperatures. The development of crazing is reduced. The first three saltings are made with sodium chloride alone, followed by firing with a mixture of salt with an addition of 15% borax, and finally by a richer borax mixture.

Boric acid (H_3BO_3). Being soluble in water, it is never used as a major glaze component, only as a frit component. It is stable at ordinary temperatures, but ideally storage should be in dry, warm conditions. When heated it gradually loses water until anhydrous boric oxide (B_2O_3) is formed. Boric acid is the preferred compound when fritting low alkali compositions. It is available in crystal, granular and powdered form. Boric oxide in the fused condition melts at about 325°C and is pourable at about 500°C. Because of this property, fused boric oxide finds a ready use in fritting where swelling of the batch during melting is unacceptable.

Borax ($Na_2B_4O_7.10H_2O$). Sodium tetraborate decahydrate is stable under normal storage conditions, but when exposed to dry air or elevated temperatures it tends to lose water of crystallisation. On heating, borax begins to melt at 62°C with frothing and on reaching a temperature of 100°C, five molecules of water of crystallisation have been lost. At 160°C four more molecules are lost until finally at 400°C it is completely dehydrated. Several differently sized grades are available.

The stable form, sodium tetraborate pentahydrate ($Na_2B_4O_7.5H_2O$), is often used because of its compatability with other raw materials which might pick up moisture and so create caking problems. The reduced content of water of crystallisation gives savings in storage and handling.

Dehydrated borax ($Na_2B_4O_7$). Sodium tetraborate is prepared by dehydrating borax. It is a hard glassy granular powder and is the preferred grade for most vitreous production processes because it neither foams nor froths. In normal ambient conditions it absorbs water and will eventually revert to a hydrated crystal form. Storage should be under cover in waterproof containers. Melting at 741°C means it is an ideal flux during the initial stages of fritting.

Colemanite ($CaB_3O_4(OH)_3.H_2O$). This mineral is a hydrated calcium borate and is almost insoluble in water. It is a powerful flux,[70] melting at about 900°C. However, in the natural form its use in glaze (as a mill addition) is limited by the ease with which it decrepitates. Precalcination reduces the violence of dehydration during early parts of the glost cycle and inclusion in raw glaze recipes becomes possible.

3.2.10 Zinc

Zinc is a valuable auxiliary flux in glazes firing up to about 1050°C. At higher temperatures it is too active in reducing the viscosity and the amount added to a glaze recipe should be adjusted to needs. By its use the firing range of high-temperature glazes is extended. Both the opacity and the whiteness of opaque glazes are improved, although it cannot be classed as an opacifier. It reduces the coefficient of thermal expansion and improves the chemical durability of some compositions.

As some glazes mature, zinc oxide will dissolve readily even if present in excessive amounts. On slow cooling, zinc silicate (Zn_2SiO_4) crystallises as if from a saturated solution. The fine crystals collect any pigment or colouring ion present in the glaze to give highly decorative effects. Precalcined with china clay, zinc oxide forms a more controllable matting agent, which when added to a suitable glaze produces a mass of small crystals and a finer surface texture than with zinc oxide alone. Other modifiers for the surface texture of zinc matts are titania and tin oxide.

The presence of zinc in a glaze can be antagonistic to some ceramic colours. It reacts with some colouring elements, notably chromium and iron, in whatever crystal form they are incorporated in the glaze, causing complete colour changes. Underglaze decoration is affected by the presence of zinc in the covering glaze. Several countries include zinc in their regulations controlling toxic metal release.

Zinc oxide (ZnO). Zinc oxide forms the basis of Bristol glazes. This is the principal source of zinc in frit and glaze manufacture. It is usually produced by the direct oxidation of zinc metal or a high zinc alloy. There are a number of different qualities available, classified according to natural colour and levels of iron, copper and lead. Calcined zinc oxide is often used where there is a need to minimise shrinkage during drying and firing,[71] shrinkage which would otherwise lead to "crawling".

3.2.11 Tin

In glaze only one tin compound, tin (IV) oxide, is commercially exploited. It has long been used as a glaze component and it was of major importance in the original majolica glazes of the sixteenth century. It is as an opacifying agent in high-quality glazes and as a colour stabiliser in some pigments or glazes that tin oxide finds a modern role. Minor quantities are used in the conducting phase in some electrical porcelian glazes.

Though of low solubility in glaze by a degree defined by composition and firing temperature, anhydrous tin oxide has little effect on the glaze surface. A small quantity only dissolves in the glaze and the physical properties of a glaze are largely unaffected. Although it has a comparatively low refractive index (within the range 1.99–2.09), tin oxide will produce high gloss white glazes. The amount needed to reach full opacity will depend upon the composition of the base glaze and upon the particle size of the tin oxide. Grades having low particle size and low bulk density together give high-quality "blue" whites, different in tone and texture from zircon opacified glaze. Curiously the addition of some fine ground zircon to a tin glaze enhances the level of opacification. High-density tin oxide is used in coloured glaze. Between 5% and 10% will opacify most transparent glazes, though some compositions might require as much as 15%.

Because as an opacifier tin oxide remains in suspension in the vitreous matrix, and does not dissolve and precipitate, the opacity of glazes containing it is not affected by the firing cycle. However, there are adverse reactions because of the ease with which highly coloured compounds can be formed with tin oxide as a base. It readily absorbs volatile chromophores such as vanadium or chromium if they exist in the kiln environment to create undesirable "flashed" areas of colour.

Tin oxide (SnO_2). Although there is only one source of tin for use in glaze, tin oxide, there are many qualities with some being more appropriate for use in pigments than as a glaze opacifier. The normal route of preparation is from the ore cassiterite, but alternative sources are obtained as a by-product in the purification of lead–tin alloys and in the scrap recovery of tin plate. Two principal types of tin oxide are available, each having markedly individual properties, appropriate either to use in glaze or pigments.

3.2.12 Lead

Lead oxide has long been a valuable component of many glazes ranging from those in the lowest melting range to those of the highest quality. Lead is a valuable network-modifier, although in high lead glasses lead ions may be regarded as taking part in the network.[72] With the lead atoms acting as bridges between isolated SiO_4 tetrahedra, its strong fluxing action allows low firing glazes to be compounded without risk from the development of high coefficients of thermal expansion when using alternative alkaline fluxes. A special type of high solubility, low melting point glaze containing 90% PbO (by weight) can be made.[73,74] With silicon as the principal network-former, glasses containing higher proportions of PbO than the orthosilicate $2PbO.SiO_2$ are possible.[75]

Glazes containing lead are characterised by their low surface tension, low viscosity, wide firing range, high index of refraction and resistance to chipping. Although they are resistant to the development of crystallisation, lead glasses readily phase separate and this property is of great significance when formulating low solubility glazes and when making glazes to conform to the international regulations for low lead release (Section 9.5). Glazes based on high lead contents have excellent solvent power for metallic oxide colorants.

The low surface tension of molten lead glaze is an important property which, in combination with the inherent long-firing range, gives highly glossed surfaces. This property of good wetting power enables lead glaze to wet and flow over surface defects in the substrate.

When substituted for alkali, there is a reduction in the coefficient of thermal expansion. While the addition of lead has little effect on elasticity,[76] there is a recognised ability of glazes high in lead to resist crazing when a mismatch of thermal expansion exists.

Glazes containing lead are used principally below 1150°C to minimise volatilisation which occurs significantly at higher temperatures, with lead compounds condensing on the cooler zones of the kiln as the temperature falls. The volatilisation losses of PbO are dependent on time and temperature and upon the level of reducing gases in the furnace. In fritting, the grain size of the source of lead oxide has a bearing on the volatility of batch components.[77]

The toxic nature of lead compounds is well known and guidance for their safe use in ceramic manufacture is codified in legislation. Raw lead glaze, that is glaze made from lead compounds other than in a stable glassy form, should not be used due to its solubility in stomach acids. Regulations covering the use of lead are given in the "Control of Lead at Work" SI.1980 No 1248 revised 1985.

Red lead oxide (Pb_3O_4). Lead orthoplumbate is used in fritting as a relible means of introducing lead with a low risk of reduction. Red lead was once used as a basis for raw lead glaze, **but should not be so employed now.** As commercially available, it averages 99.7% purity, though it contains a small proportion of lead monoxide (PbO). The mean particle size is extremely low and it is insoluble in water.

For frit production in the ceramic industry it is available in a low dusting condition with either an addition of 5% moisture or of 0.5% oil. Only traces of the major colouring elements are present and the iron content is approximately

0.0002% Fe_2O_3. Decomposition begins at 538°C with the release of oxygen, providing an important protection against further reduction to lead metal during the early stages of melting frit.

It is stable in storage.

Litharge (PbO). Lead monoxide, a major raw material for frit manufacture, occurs in two polymorphic forms, one red and the other yellow, either grade containing only very low levels of the major colouring elements.

The iron content is approximately 0.0005% Fe_2O_3. Commercial grades have a low particle size, being substantially less than 14 μm in diameter. There is a slight solubility in water. It initiates melting reactions at 888°C. Unfortunately it absorbs carbon dioxide from the air. Consequently it must be stored in dry conditions in unopened packages.

White lead ($2PbCO_3.Pb(OH)_2$). Basic lead carbonate was once the basis of raw lead glaze. It has no use in the safe and hygienic processes of modern frit or glaze manufacture due to its solubility in stomach acids.

Lead sulphide (PbS), Galena. Galena is a dense metallic-grey coloured mineral used as a basis of glazes from medieval times to the present day. There is comparative safety in its use because of its low solubility when finely ground, but the composition of the glaze it produces on the ware has variable properties. Traditional English slipware glaze was made from 3 parts galena and 1 part red clay and the firing temperature varied between 900°C and 1100°C. When glazes of this type are used it is difficult to avoid "sulphuring".

Lead silicate. Now manufactured in several grades by suppliers of lead compounds, lead "mono"-silicate is an alternative and safe source of lead for melting lead borosilicate frits. By its use as a replacement for lead oxide powders, "lead-in-air" values during batch mixing can be reduced by as much as 90%. In the lead silicates available, the lead content covers the range from 65% PbO to 85% PbO in weight percent. They are not classed as "low solubility" frit and are only sparingly used as a mill addition during the grinding of glaze. The glassy lead silicates (nominally $PbO.SiO_2$ and $3PbO.2SiO_2$) are supplied in granular form and with their use there is less mechanical loss of lead, through entrainment in furnace gases, when charging the frit furnace. In this form lead is less volatile; therefore, there is a lower loss of lead through volatilisation at high temperatures. Additionally corrosion of furnace refractories is reduced, giving improved quality to the final glaze.

Commercial grades are available with very low iron contents between 0.005% and 0.015% Fe_2O_3 and with moisture contents of between 0.5% and 2.0% H_2O. The melting range of these silicates is 677–732°C.

Lead silicate ($3PbO.SiO_2$). Called tri-basic lead silicate, this glass has the highest lead content, 92% (by weight) PbO, of any fritted silicate. It is widely used for the introduction of lead to many types of vitreous composition. It is available in a coarse granular form which helps to reduce loss of lead during frit melting through entrainment in furnace gases or by volatilisation. The aluminium content is deliberately kept as low as possible (about 0.006% Al_2O_3) to maintain the excellent solvent action for other batch materials. The iron content of the tribasic lead silicate is about 0.004% Fe_2O_3 and other colouring metal ions are at a very low level. It melts in the range 677–732°C.

As an additional aid to hygiene, lead silicate is available containing 2% moisture.

Lead bisilicate ($PbO.2SiO_2$). Lead bisilicate is one of the most important raw materials for glaze as a source of lead oxide either in the batch when making frit or when ground as a component of the glaze.

Phased increases in the lead oxide content in the $PbO–SiO_2$ glass-forming system are not matched by any corresponding increase in the solubility of the glass in acid solutions, but rather are there alternate rises and falls in the resistance to acid attack.[78,79] The lowest lead solubility occurs in the region of the compositions $PbO.2SiO_2$ and $PbO.3SiO_2$. Lead bisilicate contains 65wt% PbO and 32.2wt% SiO_2, completely in the glassy form with a small addition of aluminium oxide for extra chemical resistance. Even when waterground it has only a low solubility in many acids (including body fluids) and when correctly melted it is recommended as a safe source of lead in all glaze compositions.

Novel preparation methods eliminate as far as possible contamination by the common colouring elements. The iron content is very low at 0.025% Fe_2O_3 and the frit can be used in white pastel and highly coloured glazes. It is supplied as pale clear lemon yellow granules slightly dampened for additional safety.

Note: When using any lead compound the adoption of correct handling methods is essential. By following simple rules the safe and hygienic handling of lead chemicals and frits is easily achieved. Guidance on these matters is usually available from the suppliers of lead compounds or from Statutory Instrument 1980 No. 1248.

3.2.13 Colorants in Ceramic Glazes

These can be classed as additives or as raw materials. Several simple compounds can be used as components of frit or as mill additions, whilst others are best added for effect or economy only at a later stage of glaze preparation.

Colour is produced through the absorption of characteristic wavelengths either by elements dissolved in the glaze or by inert pigment particles suspended in the glaze. The absorption spectra of transition metal ions and rare earth ions dissolved in glass are of interest because of the information they give to investigators about structure.[80] Certain absorption peaks of colours in glass are similar to those of known metal ions in crystals and in aqueous solutions and it has been deduced that the same valency states exist in each. The absorption bands are related to electron transitions between different levels in one ("d" or "f") shell of the ion. There are two main and one secondary method of introducing colorants in glaze.

Transition metal compounds. When dissolved in a glaze or frit, the colour which develops is due to the influence of neighbouring ions and to the metal ions' own valency state. Nearby ions can alter the electronic energy levels of the metal ions, thereby influencing the absorption spectra. Altering the composition of the solvent glass will usually alter the colour perceived by the eye and with the same metal compound a leadless (basic) glaze and a lead (acid) glaze can produce two different colours.

From the first transition metal group the glaze technologist uses chromium, cobalt, iron, manganese, nickel, copper, vanadium and titanium. In the later lanthanide group there are useful colorants such as cerium, praseodymium and neodymium.

Ceramic pigments.[81] Specialist inorganic pigment manufacturers have developed the technology of "colour" making to a high degree. Glaze stains are prepared from pure raw materials, which are carefully mixed before calcination at precise temperatures. Subsequently the calcined cake might be ground to a powder in any of several ways. Soluble salts which might be present can be neutralised, precipitated or removed by washing.

Most ceramic stains change in intensity if they are ground past the optimum particle size. Consequently, they should not be added to the glaze mill at too early a stage. Ideally the stain should be ground only until it is evenly distributed in the glaze and unlikely to give coloured specks.

Coloured frits. Though not widely used, they have useful characteristics in some applications, particularly in transparent glazes. The technology is one of coloured glasses.[82,83] The frits contain transition metal compounds dissolved in a highly solvent glass, consequently the coloration of the final glaze to which they are added is very even. Gas evolution normally associated with the interaction of the compounds of some elements (e.g. manganese dioxide) with glaze components is completely absent and blisters or bubbles are consequently rare.

The colour of a glaze is influenced by the colour of the ceramic body on which it is fired, the amount of pigment, the way the glaze is applied, the composition of the glaze, the temperature to which it is fired, the rate of temperature rise and the furnace atmosphere.

Chromium

Chromium appeared as a glaze colorant at the beginning of the nineteenth century as a substitute for the green produced by copper. In most compositions the colour developed by chromium is a characteristic green, but in high lead glazes a nasturtium red develops at low firing temperatures, probably due to the formation of lead chromate.

Many different compounds can supply chromium ions, e.g. chromic oxide

(Cr_2O_3) or lead chromate ($PbCrO_4$). The solubility of chromium varies according to the frit or glaze composition and the firing temperature. Chromium solution colours are improved by the addition of boric oxide to the frit batch. Glazes based on lithium are better solvents than those based on soda or potash. As the temperature of a fluid glaze coloured with chromium decreases from a peak, spangles of chromium oxide crystals separate out in an aventurine effect.

Zinc and tin produce characteristic side effects in the presence of chromium. Zinc and chromium combine to produce dull brown colours. Small quantities of chromium oxide (1%) combined with a tin oxide opacified glaze produces a pink.

Above 1100°C chromium compounds in glaze become volatile and the effects of flashing of the colour onto adjacent pottery is often seen.

Chromium oxide (Cr_2O_3). Green chromium (III) oxide is the most stable oxide and it is available as a high-quality fine particle sized powder, which is insoluble in water. It is not very soluble in glaze, except in those with high alkali and high B_2O_3 contents. Normal additions lie between 2% and 5% and at these levels glazes are opaque.

Iron chromite. "Chrome ore", iron chromite, is used as a source of grey brown specks in all types of glaze. As a finely ground powder it is a stable grey pigment under many different kiln atmospheres.

Cobalt

Although cobalt was isolated only in the eighteenth century its use as a glass colorant has been known for about 4500 years. Materials having cobalt as their base were amongst the first examples of synthesised pigments. The Portland vase, which was imitated in ceramics by Wedgwood, is coloured dark blue by cobalt. The oxide belongs to the RO group.

Nearly all glasses develop a blue colour with cobalt (as Co^{2+}), although in some compositions a pink will develop. The solubility of cobalt salts in glass is high. Cobalt is a powerful colorant and as little as 0.02% CoO produces a noticeable tint in a transparent glaze. Use is made of the powerful colouring property to mask the undesirable yellow tones of some lead glazes and some borax frits when about 0.005–0.01% CoO is required. At about 0.2% CoO the glaze will have a definite blue colour. Because of the intensity of its coloration, control of the depth of colour is difficult when using cobalt oxide. In the making of transparent blue glazes, prior dilution in a frit or base stain is a requisite.

Glasses containing up to 12% CoO can be melted,[84] but so intense is the colour that the glass appears black. Crystalline glazes containing up to 10% CoO remain blue. Bodies and engobes can carry more cobalt than glaze without going black. Cobalt oxide is an essential component of the best black ceramic pigments.

The blue colour resulting from the use of cobalt is unaffected by changes in the firing schedule or atmosphere. Normally cobalt is not volatile even at the highest glazing temperatures (1400°C), but in the presence of chloride ions it is volatile and a blue coloration will strike onto other nearby glazed surfaces. At any level of addition, soft low firing leadless and low solublity lead glazes give the brightest colour.

In glazes containing large amounts of magnesia, cobalt produces a pink–lavender mauve colour. The pink colour can be enhanced in alumina-free glazes.

Cobalt oxide (Co_3O_4). Most commercial grades containing about 71–72% Co are mixtures of two oxides Co_3O_4 and CoO. It is a uniform, black fine powder, free from grit. The particle size range is narrow and 100% of the material is finer than 20 μm. For ceramic use control is exercised over the contents of iron, manganese and copper so that bright sharp blue tones develop.

One per cent in a glaze gives a strong blue, whilst at a 4% addition the blue takes on distinct black tones.

Cobalt carbonate ($CoCO_3$). Commercial salts containing about 47% (by weight) Co are often mixtures of the basic carbonate, carbonate and hydroxide. Because it has only two-thirds of the level of cobalt it is easier to use than black cobalt oxide. The low level of metal impurities are responsible for the clean blue tones which develop. Conversion, on heating, to cobalt oxide is easily achieved, so faults due to the evolution of gas even in low melting point glazes are not a problem. It can be used at levels up to about 1.5%.

Cobalt sulphate ($CoSO_4.7H_2O$). A decorative effect on glaze with a solution of cobalt sulphate can be developed either on the unfired or the fired glaze. When sprayed on the unfired glaze the solution is absorbed to give a uniform pale blue at glost temperatures. If a solution of cobalt sulphate is allowed to crystallise on the already fired glaze[85] the crystals can be fixed "in-glaze" by firing the ware to between glost and enamel temperature.

Smalt and zaffre.[86] These old terms do not represent precise compositions and modern standardised equivalents are available, either as prepared cobalt aluminate or cobalt silicate stains.

Copper[87, 88]

As one of the strongest colouring elements, copper was one of the earliest colorants of glaze in early Egyptian faience. It gives a wide range of colours from blue to green and in special circumstances "red" depending upon composition and upon the firing atmosphere. Any other shades seen in copper glazes are usually the result of the development of these colours at the same time within the glaze. Copper compounds are very soluble in all types of glaze or frit.

The colour of any glaze containing copper will depend upon the amount of copper present, the kiln atmosphere (whether oxidising or reducing), which alkali is present, and whether lead oxide or tin oxide is present. To generalise, alkaline glazes produce blue colours and acid glazes develop a green colour with copper. In order to maintain the colour in a frit an oxidising agent should be added to the batch. Borax in the frit keeps the colour bright. In a lead glaze 1–6% CuO will produce shades of apple or grass green. By increasing an addition of TiO_2 to the base glaze the colour can be moved through green to brown. When iron oxide is present in appreciable amounts the colour develops with an unattractive greenish tint.

Coloured crystals can be developed by slowly cooling glazes high in zinc oxide and calcium oxide in which 1% copper oxide has been dissolved. An excess of copper beyond its natural solubility in the glaze (about 5%) will give a metallic grey surface.

Copper glazes are easily reduced and the developed colour quickly responds to changes in atmosphere. A green glaze can be converted to a red glaze in a reducing atmosphere. The brightest reds develop at low concentrations of copper under the influence of heavy reduction of the glaze. For this effect the ideal glaze base is a low melting point composition containing between 0.2–8% CuO. Tin oxide (about 1%) is an essential secondary ingredient. After an oxidising fire to take the copper compound into solution a reduction phase is introduced into the cooling cycle during the time the glaze is at its most responsive—150°C above its softening point.

Copper is an active flux in glaze and it will reduce the firing range of any composition to which it is added. It should never be used in glazes for any food contact surfaces.

Copper oxide (CuO). Black cupric oxide is the main source of copper in glaze and it can be used as an addition to a raw glaze or dissolved in a frit. It can be painted directly onto unfired biscuit glaze and allowed to react in the glost fire. Painted under the glaze, copper oxide behaves as a reactive colour.

Copper carbonate ($CuCO_3.Cu(OH)_2$). Known as malachite green, basic copper carbonate is used to introduce copper into those glazes which will be subjected to reduction. It is insoluble in water and can be added to the mill. In a raw glaze containing tin oxide and fine silicon carbide, copper carbonate will give purple red colours. Where freedom from the risk of speckle (which is sometimes associated with cupric oxide additions) is required copper carbonate is the preferred additive.

Iron

Iron compounds are present in all ceramic raw materials. As a colorant, iron reacts in a variable way, the effect produced depending on the glaze composition and the atmosphere in which it is fired. The "life" in old glazes is due in part to the presence of iron (and other elements) in the raw materials of which the potters of those days were not aware.

In some borax frits the colour from iron in solution is blue, but more usually the colour is yellow to yellow–green. In lead borosilicate frits the colour is red-brown. As a general rule under oxidising conditions the iron colour will be reddish; in reducing conditions the colour is grey and there is a range of yellows, greens and blues in between. In glazes saturated with iron oxide (about 12%) the colours are dulled, but in the presence of titania, and with a slow cooling rate, aventurine effects develop. When added to a mill the range is from 1% of iron oxide which gives a slight coloration up to 12% Fe_2O_3 as a maximum, but at this high level the surface is affected and "mottled" colours develop.

Iron dissolved in the glaze is easily reduced and in this condition it is the basis for an important series of art glaze effects (celadon, tenmoku, tessha and kaki). Reduction of the iron is most effective immediately before the glaze begins to melt.

Modifiers for the colours developed by iron are zinc which produces green tones, titanium which produces yellows and orange and tin oxide with which iron develops cream–coffee colours. Lime has a pronounced bleaching action on the developing colour of iron stained glazes. Iron oxide is used to modify and mellow other colours.

Iron oxides (Fe_2O_3, Fe_3O_4). There are many grades available with all levels of purity and particle size. The natural colours are from red to black. Some grades ("black iron oxide") act as fluxes whilst the behaviour of other grades is either as a flux or as a refractory.

Black iron oxide is used in speckled glazes.

Iron hydroxide (α–FeOOH). "Yellow iron oxide" is a fine powder which can colour glaze uniformly without specks.

Manganese

Managanese exists in a number of valencies and there are a number of derived colours. The tone of these weak colours in either glaze or frit can be modified by changes in composition.

In oxidised glazes the colour of manganese is pink, but in a glaze fired in a reducing atmosphere, deep greenish-brown colours predominate. As the "oxide", up to 4% will readily dissolve in most glazes, but thereafter precipitation as crystals increasingly occurs on the glaze surface. From 5% to 10% produces shades of purplish brown and, at 20%, the surface (though dark brown) is metallic. In crystalline glazes it produces clusters of coloured crystals. Overfiring causes managanese solution colours to fade.

Manganese compounds act as fluxes and they increase the fluidity at lower melting temperatures. It is an important colorant in Rockingham-type glaze in which it is associated with iron oxide in a transparent lead glaze base. The rich brown colour can be modified with occasional crystalline striations, whose development is related to the firing conditions.

Manganese dioxide (MnO_2). Known once as magnesia nigra, manganese dioxide or pyrolusite has been used in glazes to give mainly brown colours; however, in a high potash glaze attractive purple colours develop.

Many qualities are available containing from 40% Mn to 65% Mn. Before adding to a glaze, manganese dioxide is best calcined because at 500°C it decomposes, losing oxygen. In glaze there is further reaction when as MnO it reacts at 700°C with alkali and oxygen to form manganates. At 900°C this ceases and the balance between Mn(II) and Mn(III) reaches equilibrium. These changes give rise to bubble and blister formation.

Granular manganese dioxide gives a speckled effect in low temperature glaze, but as the firing temperature rises, the grains dissolve erratically and the effect is variable.

Manganese carbonate ($MnCO_3$). The very fine particle size of manganese carbonate makes this a source of manganese when freedom from glaze specking is desirable. It is insoluble in water and can be added directly to the mill. Although it loses CO_2 on heating, it is not so prone to produce blistering as is pyrolusite.

Nickel

Nickel when dissolved in a frit is a powerful colorant—as little as 0.002% Ni gives a noticeable tint. The normal additions are from 2% to 5% nickel oxide. When in complete solution in a frit or transparent glaze the colours which develop include yellow, brown, blue and pink, depending upon the composition.[89] In a high potash glaze and high lead glaze the colour is pink, while in a high soda glaze the colour is brown. In a lithium glaze the colour is yellow. In high zinc glaze the colour which develops is increasingly blue.[90] BaO and CaO in the glaze favours a brown. Variation in the glaze thickness affects the development of most nickel colours.

It is rarely used in low-temperature glazes.

Nickel oxide (NiO). Nickel oxide is a grey–green, high purity, finely ground powder containing 77–78% Ni. Grades for ceramic use have very low levels of other colouring elements. It is insoluble in water and can be added directly to the mill to produce colours ranging from yellow to brown, green to blue or grey according to the chosen composition of the base glaze.

Nickel oxide is refractory and it will increase the maturing temperature of a glaze.

Nickel carbonate ($NiCO_3x.Ni(OH)_2$). Basic nickel carbonates are obtained from the interaction of alkali metal carbonates and nickel salt solutions. They are light green powders which can be used directly as mill additions. Decomposition of the carbonate is rapid when heated, leading to the formation of nickel oxide.

Vanadium

Vanadium has a comparatively weak colouring effect in solution either in frit

TABLE 3.1. METAL OXIDE COLOUR DEVELOPMENT IN GLAZE

	% addition	Colour in "lead" glaze	Colour in leadless glaze	Colour when fired in reducing atmosphere
Chromium oxide	2%	Vermilion at 900°C (very high lead) Green at 1050°C	Yellowish green	Emerald green
Cobalt oxide	0.5%	Medium blue	Medium blue	Medium blue
	1%	Strong blue	Strong blue	Strong blue
Copper oxide ("black")	0.5%			Copper red
	1%	Green	Turquoise	Deep red
	2%	Deep green	Turquoise	Red and black
	8%	Green with metallic areas	Blue–green with metallic areas	Black
Iron chromite	2%	Grey–brown	Grey	Grey
Iron oxide	1%			Celadon
	2%	Yellow amber	Tan	Olive green or celadon
	4%	Red–brown	Brown	Mottled green
	10%	Dark red	Black–brown	Saturated iron red brown
Manganese dioxide	4%	Purple-brown	Purple	Brown
Nickel oxide	2%	Grey–brown	Grey	Grey–blue
Rutile	5%	Tan	Grey–brown	
Vanadium pentoxide	6%	Yellow	Yellow	
Cobalt oxide Iron oxide	0.5% 2%	Grey blue	Grey blue	Grey green
Cobalt oxide Manganese carbonate	0.5% 4%	Blue-purple	Aubergine	Blue
Cobalt oxide Rutile	0.5% 3%	Grey blue	Grey blue	Crystalline blue break-up
Copper carbonate Rutile	3% 3%	Crystalline green	Crystalline blue–green	
Iron oxide Rutile	2% 2%	Crystalline brown	Crystalline gray–brown	Crystalline brown break-up
Iron oxide Cobalt oxide Manganese carbonate	8% 1% 3%	Black	Black	Black
Cobalt oxide Iron oxide Manganese carbonate	3% 2% 2%	Mirror black	Black	Black
Manganese dioxide Iron oxide	6% 3%	"Rockingham" brown		
Cerium oxide Titania	2.5% 2.5%		Yellow	Yellow
Neodymium oxide	3.0%	Violet	Violet	Violet

or glaze and it is rarely used as a pigment. More often it is used as a component of stable ceramic stains.

It is a very active flux (melting at 690°C) and it reduces the surface tension and fluidity of a glaze at the peak firing temperature.

Low concentration in a lead glaze (about 1%) gives a pleasing yellow colour, but this changes to a muddy brown as the level of vanadium increases to about 10%. These high concentrations are usually associated with a patchy, textured, effect. Greener colours develop if the glaze is fired in reducing conditions.

Vanadium pentoxide (V_2O_5). Vanadium pentoxide which is sparingly soluble in water can be used as a mill addition in the preparation of yellow or brown reactive glazes. Used in conjunction with tin oxide clean yellow colours form at most glaze firing temperatures. It is harmful.

Ammonium metavanadate (NH_4VO_3). An alternative source of vanadium is the precipitated salt ammonium metavanadate. When used as a mill addition it is the source of much reactivity in the molten glaze. It dissociates at 210°C.

REFERENCES

1. Worrall, W. E. *Ceramic Raw Materials*, 2nd ed. Pergamon (1982).
2. *Glass Making Today* (Ed. P. J. Doyle) pp. 2–9. Portcullis Press (1979).
3. Ramsden, C. E. and Ryder, S. H. "Batch materials excluding sand and soda ash". *J. Soc. Glass Tech.* **40,** 388–403 (1956).
4. Potts, J. C., Brookover, G. and Burch, O. G. *J. Amer. Ceram Soc.* **27,** 225–231 (1944).
5. Mason, Brian. *Principles of Geochemistry*, pp. 100–104. John Wiley & Sons (1966).
6. Worrall, W. E. *Raw Materials*, Institute of Ceramics Textbooks, p. 15. Maclaren (1964)
7. Smith, Martin. "Loch Aline Sand". *Ind. Minerals*, p. 67 (1981).
8. *Glass Making Today* (Ed. P. J. Doyle), pp. 2–9. Portcullis Press (1979).
9. Anon. "Flint". *Ind. Min.* pp. 21–29 (1979).
10. Clarke, Gerry. "Cheshire Silica Sand". *Ind. Minerals*, p. 28 (1982).
11. Watson, Ian. "Nilsea quartz". *Ind. Minerals*, p. 28 (1982).
12. Batchelor, R. W. *Flotation Felspar. Trans. Brit. Ceram. Soc.* **72,** 7–10 (1973)
13. Segrove, H. D. *J. Soc. Glass Tech.* **40,** 363–375 (1956).
14. Anon. *Glass*, **52,** 273–276 (1975).
15. Taggart, A. F. *Handbook of Mineral Dressing*, 11.62–11.65. John Wiley (1954).
16. Taggart, A. F. *Handbook of Mineral Dressing*, 12.125–12.128. John Wiley (1954).
17. Rawson, H. *Proceedings IVth International Congress on Glass, Paris.* pp. 62–69 (1956).
18. Mason, Brian. *Principles of Geochemistry* John Wiley & Sons (1966).
19. Evans, K. A. and Brown, N. "Speciality Inorganic Aluminium Compounds" pp. 164–195. *Speciality Inorganic Chemicals* (Ed. Thompson, R.) (1981).
20. Dimbleby, V., Hodkin, F. W. and Turner, W. E. S. "The influence of aluminium on properties of glass". *J. Soc. Glass Technol.* **5,** 107–115 (1921).
21. Parmelee, C. W. and Harman, C. G. *J. Amer. Cer. Soc.* **20,** 224–230 (1937).
22. MacMillan, P. W. Glass Ceramics, pp. 14–15 2nd Edition. Academic Press (1979).
23. Holloway, D. G. *The Physical Properties of Glass.* pp. 16–18. Wykeham Publications (1973).
24. Coope, B. M. "Kaolin". *Ind. Minerals,* No. 136, pp. 31–49 (1979).
25. Worrall, W. E. *Ceramic Raw Materials*, 2nd ed. pp. 56–68. Pergamon (1982).
26. Firth, E. M. Hodkin, F. W., Parkin, M. and Turner, W. E. S. *J. Soc. Glass Technol.*, **10,** 365–368 (1926).
27. Taylor, J. R. *Glass Technol.* **23,** 251–261 (1982).
28. Batchelor, R. W. *Flotation Felspar. Trans. Brit. Ceram. Soc.* **72,** 7–10 (1973).
29. Royle, J. D. *Trans. Brit. Ceram. Soc.* **73,** 291–296 (1974).

30. Lyng, S. and Gamlem, K. *Trans. Brit. Ceram. Soc.* **73,** 133–142 (1974).
31. Royle, J. D. *Trans. Brit. Ceram. Soc.* **73,** 291–296 (1974).
32. Charles, R. J. *J. Amer. Ceram. Soc.* **48,** 432 (1965).
33. Dietzel, A. H. *Physics and Chemistry of Glass,* **24,** 172–180 (1983).
34. Hummel, F. A. *Thermal Expansion Characteristics of Natural Lithia Minerals.* Foote Prints **11** (1948).
35. Lester, W. R. *Cent. Glass Ceram. Res. Inst. Bull.* **1,** 34–37 (1960).
36. Koch, W., Harmons, C. and Bannon, L. O. "Some physical and chemical properties of experimental glazes". *J. Amer. Ceram. Soc.* **33,** 102–107 (1950).
37. Dietzel, A. H. *Physics and Chemistry of Glass,* **24,** 172–180 (1983).
38. Laidler, D. S., *Lithium and Its Compounds.* Monograph of Royal Institute of Chemistry (1957).
39. Ed. Thompson, R. *Speciality Chemicals,* Royal Society of Chemistry 1981 Chapter on "Lithium Chemicals" pp. 98–122 by J. E. Lloyd.
40. Richardson, F. W. "Use of lithium carbonate in raw alkaline glazes". *J. Amer. Ceram. Soc.* **22,** 50–53 (1939).
41. Simmingsköld, B. and Jönsson, B. *Glastek. Tidskr.* **15,** 43–47 (1960).
42. Suppliers data.
43. Firth, E. M., Hodkin, F. W., Parkin, M. and Turner, W. E. S. *J. Soc. Glass. Tech.* **10,** 368–370 (1926).
44. Ed. Doyle, P. J. *Glass Making Today* p. 169 Portcullis Press (1979).
45. Anon., *Verre Silic. Ind.* **5,** 205–208 (1934).
46. BS 3108:1980 Limestone for making colourless glasses (Amendment AMD3600:1981).
47. Hodkin, F. W. and Turner, W. E. S. *J. Soc. Glass. Tech.* **5,** 189–194 (1921).
48. Kreidl, N. J. *The Possibility of Strontia in Ceramics.* Foote Prints **14,** p. 24 (1941).
49. Ramdsen, C. E. "A study of the chromium red glaze". *Trans. Eng. Ceram. Soc.* **12,** 239 (1912–13).
50. Koening, C. J. "Nepheline Syenite in low temperature glaze". *Eng. Expert. Station Bull.* Ohio State University pp. 33–36, No. 112 (1942).
51. McCutcheon, E. S. *J. Amer. Ceram. Soc.* **27,** 233–238 (1944).
52. Harman, C. G. & Swift, H. R. "Raw leadless white glazes". *J. Amer. Ceram. Soc.* **28,** 48–52, (1945).
53. Gray, T. J. "Strontium glazes and pigments", *Bull. Amer. Ceram. Soc.* **58,** 768–770 (1979).
54. Shteinberg, Ya. G. *Strontium glazes.* Trans. T. R. Gray (1974).
55. Ed. Thompson, R. *Speciality Inorganic Chemicals* (1981) Chapter "Magnesium Compounds" pp. 123–163. T. P. Whaley.
56. Ed. Thompson, R. *Speciality Inorganic Chemicals,* Royal Society of Chemistry (1981) Chapter "Inorganic Titanium compounds" by Eveson, G. F. pp. 226–247.
57. Hepplewhite, J. W. *J. Amer. Ceram. Soc.* **20,** 66–61 (1937).
58. Taggart, A. F. *Handbook of Mineral Dressing,* pp. 11–104 to pp. 11–125. John Wiley (1954)
59. Taggart, A. F. *Handbook of Mineral Dressing,* pp. 12–125 to pp. 12–128. John Wiley (1954).
60. "Enamelling of sheet iron hollow ware". *Keramische Rundshau,* **16,** 89, 135 (1908).
61. Booth, F. T. and Peel, G. N. *Trans. Brit. Ceram. Soc.* **58,** 532–563 (1959).
62. Hepplewhite, J. W. *J. Amer. Ceram. Soc.* **20,** 60–61 (1937).
63. Taylor, W. C., B. P. 118397 (1918), 118398 (1918).
64. Rawson, H. *Inorganic Glass Forming Systems.* p. 235. Academic Press (1967)
65. Callow, R. J. *J. Soc. Glass Tech.* **36,** 266–269, 270–274 (1952).
66. Neri, Antonio. *L'Arte Vitraria,* published in Florence in 1612. An English version translated by Christopher Merrett was issued in London in 1662.
67. Weyl, W. A. Boric Oxide, Its Chemistry, *Technical Service Bull. No. 13,* Borax Consolidated Ltd.
68. Firth, E. M., Hodkin, F. W., Parkin, M. and Turner, W. E. S. *J. Soc. Glass Tech.* **10,** 370–373 (1926).
69. Weyl, W. A. "Boric oxide, its chemistry" *Technical Service Bulletin No. 13.* Borax Consolidated Ltd.
70. French, Myrtle M. *J. Amer. Ceram. Soc.* **14,** 730–741 (1931).
71. Reising, J. A. *Bull. Amer. Ceram. Soc.* **41,** 497–498 (1962).
72. Stanworth, J. E. *J. Soc. Glass Technol.* **32,** 161–166T (1948).
73. Rawson, H. *Inorganic Glass-Forming Systems* p. 104 Academic Press (1967).

74. Data from Cookson Group p.l.c.
75. McMillan, P. W. *Glass Ceramics*, 2nd edn. p. 24. Academic Press (1979).
76. Morey, G. W. *The Properties of Glass*, p. 301. Rheinhold (1938)
77. Ott, W. R., McLaren, M. G. and Hursell, W. B. *Glass Technol.* **13,** 154–160 (1972).
78. Rieke, R. and Mields, H. *Ber. Deut. Keram Ges.* **16,** 331–349 (1935).
79. El-Shamy, T. M., and Taki-Eldin, H. D. *Glass Technology,* **15,** 48–52 (1974).
80. Ed. Mackenzie, J. D. *Modern Aspects of the Vitreous-State* Vol 2 "Ligand Field Theory and absorption and spectra of Transition metal ions in glass". pp. 195–254.
81. *Institute of Ceramics Textbook.* To be published.
82. Weyl, W. A. *Coloured Glasses.* Pub. Soc. Glass Technol. (1959).
83. Paul, A. *Chemistry of Glasses.* Chapman & Hall (1982).
84. Fuwa, K. *J. Japan Ceram. Assoc.* **46,** 644–646 (1938).
85. Mellor, J. W. *Trans. Ceram. Soc.* **36,** 13–15 (1937).
86. Taylor, J. R. *Science and Archaeology*, No. 19, 3–15 (1977).
87. Mellor, J. W. *Trans. Ceram. Soc.* **35,** 364–378, 487–491 (1936).
88. Weyl, W. A. *Coloured Glasses*, pp. 424–435. Pub. Soc. Glass Technol. (1959)
89. Taylor, J. R. *J. Soc. Glass. Technol.* **43,** 29N–31N (1959).
90. Pence, F. K. *Trans. Amer. Ceram. Soc.* **14,** 143–151 (1912).

Chapter 4
Frit Preparation

4.1 THE PLACE OF FRIT IN GLAZE

Frits are indispensible constituents of most industrial glazes which mature at temperatures below 1150°C. Glazes employed in the production of twice-fired (i.e. with a biscuit and glost fire) earthenware, bone china, felspathic china and vitreous hotelware contain a major proportion of fritted components. Most tile glazes also include frits. With the advent of appreciably faster firing schedules, the use of a greater proportion of frit in glazes is to be expected, even in compositions which require peak temperatures above 1150°C, to ensure maturity of the glaze.

In many types of glaze, it is necessary to employ frits if the required surface appearance and durability charateristics of the ware are to be achieved after firing on pre-set schedules. For such glazes, the desired surface properties (e.g. gloss) are attained by introducing larger amounts of fluxing constituents into the glaze, that is as frits, than could be achieved by adding insoluble minerals or manufactured chemicals as mill components.

4.2 CONSTITUTION OF FRIT

Frits are glasses prepared by heating to a high temperature a mixture of sundry raw materials in a gas- or oil-fired furnace or in an electric melting unit. These constituents are chosen from the readily available inorganic raw materials. With a few exceptions, their crystalline components are broken down to acidic or basic oxides, which, in combination, form the vitreous glassy base of glaze.

Silica is the main acidic constituent of the more usual glaze frits and with boric oxide acting in a subsidiary role is the main glass network-forming oxide. Phosphate-based glaze frits are uncommon. In frit melting, silica is heated with "fluxes", the commonest of which are compounds of the alkali metals, sodium and potassium. These together with compounds of the alkaline earth metals calcium and magnesium are network modifiers. Barium, zinc and strontium compounds are used to some extent to replace some, or all, of the more widely used lime and magnesia. It is recognised that a multiplicity of bases in a glass is an aid to melting, to strength and to chemical durability. Because of this multiplicity it is customary, though arbitrary, to express frit and glaze compositions in the form of molar fractions with the bases, the acidic oxides and amphoteric oxides grouped together in a stylised formula, first proposed by H. A. Seger. Lead oxide is included with the base oxides. The fluxing properties of the available basic frit ingredients (in decreasing order of effectiveness) is:

Li_2O Na_2O K_2O BaO CaO SrO MgO ZnO.

The desired glost temperature and the possibility of cation release from the glaze affects the degree of usefulness of lead oxide as a flux.

The product of the heating process is a homogeneous glass which is almost insoluble in water, thereby providing the source of a major proportion of the flux needed for bringing glaze-maturing temperatures to practical levels.

4.3 USES OF FRIT

1. As a glaze component.
2. As a special bonding material in abrasives.
3. As a flux in subsidiary ceramic processes such as decorating.
4. As a coating for electronic components.
5. As a cement in ceramic/metal and glass/metal seals.

4.4 FRIT MANUFACTURE

4.4.1 Raw Material Considerations

Manufacture of frit involves accurately weighing out the constituent raw materials, blending to produce a uniform batch and heating in a refractory-lined furnace to form a homogeneous glass which is then cooled and solidified by either pouring into water or passing through water-cooled rollers resulting respectively in the formation of a "crystal" or a flake product.

Although it is not generally necessary in frit manufacture to use high-purity raw materials, because the impurities, if major, can usually be allowed for in compounding the frit batch, raw materials of consistent composition from delivery to delivery are required if frits and glazes having consistent properties are to be produced. For example, incompletely-refined sodium borates of consistent quality are available, but these contain SiO_2 as an impurity. Normally an allowance for this SiO_2 is made by reducing the level of silica sand employed or by an appropriate adjustment in the level of another silica-containing raw material.

Colouring chromophores in raw materials are undesirable. Most minerals contain at least trace quantities of Fe_2O_3 and attention must be paid to the level of iron present in frit raw materials. Silica sand, being a major ingredient of most frit batches, must have a low Fe_2O_3 content (not greater than approximately 0.03%) to prevent fritted transparent glazes having a marked and undesirable yellow colour. Raw materials containing more than a trace of TiO_2 can result in significant coloration in frits and glazes if PbO or even minor amounts of Fe_2O_3 are also present in the compositions.

There are many raw material combinations which, after the fritting process, result in a glass of identical chemical composition. For reasons of economy that combination of raw materials which ultimately leads to the lowest manufacturing cost, consistent with the product specification, will normally be employed.

For many oxide constituents of frits, there is more than one source of the oxide. Silica is, for example, the major oxide constituent of silica sand, felspars and kaolin. For reasons of economy B_2O_3 and Na_2O would be introduced as far as possible by the use of one of the crystal forms of sodium borate rather than by employing boric acid and either sodium carbonate or sodium nitrate. (As the ratio of B_2O_3:Na_2O in a frit is unlikely to match that which is in $Na_2B_4O_7$, a balancing amount of boric acid or sodium salts would usually be necessary.) Similarly, the use of kaolin or felspar to introduce Al_2O_3 into a composition would be preferred to the addition of the more expensive alumina or alumina hydrate unless purity of colour was important.

The use of more expensive sources of oxides can be justified if a final product, free from contaminating trace elements, is desired. Although sodium and potassium nitrates are expensive sources of sodium and potassium oxides, a small amount of these nitrates (1–2%) in the batch can produce cleaner coloured frits as a result of oxidation processes occurring during melting, following the release of oxygen during their decomposition. By such use of nitrates, clear white transparent borax frits can often be obtained, whereas when nitrates are not present a greenish coloured product is obtained and the additional cost of employing nitrates may be justified by this improvement in colour. Factors such as batch mixing characteristics and speed of reaction within the frit melting furnace can also lead to the use of more expensive raw materials, if this reduces the total production cost of the final fritted composition.

4.4.2 Basic Raw Materials

The following minerals and chemicals are the principal raw materials used in commercial frit manufacture:

Silica sand, SiO_2
Sodium borate, $Na_2B_4O_7$, $Na_2B_4O_7.5H_2O$ and $Na_2B_4O_7.10H_2O$ (also partially refined grades)
Boric acid, H_3BO_3
Limestone, $CaCO_3$
Felspars, soda and potash types
Kaolin (china clay)

Lead oxides (e.g. red lead oxide Pb_3O_4)
Zircon, $ZrSiO_4$
Zinc oxide, ZnO
Alkali metal carbonates and nitrates, e.g. Na_2CO_3, $NaNO_3$, K_2CO_3, KNO_3

Less commonly used materials include lithium carbonate, magnesium, strontium and barium carbonates, alumina and alumina hydrate. When both MgO and CaO are to be introduced into a frit composition, the mineral dolomite, a mixed carbonate of magnesium and calcium, provides a cheap source of the oxides.

4.4.3 Production Sequence

The first stage of the frit manufacturing process is the weighing out of the batch, which usually involves feeding major raw materials from storage silos into a weighing hopper by means of a conveyor system (e.g. belts) until the required amounts of each component have been placed in the hopper. Less commonly employed materials will be weighed out directly from bags. Due allowance must be made for any moisture associated with a raw material. Testing equipment which allows moisture contents to be rapidly determined is available. Fully computerised systems which carry out the weighing and subsequent processes prior to the actual fritting are manufactured and are particularly suitable when all raw materials can be stored in suitable silos.

After the batch has been weighed out, it is well blended ready for charging the frit kiln.

4.5 MIXING[1] BATCH FOR FRIT MELTING

The mixing operation when making frit is a two-stage process; the first (major) stage is the mixing of the dry powder as batch and the second (minor) phase is the mixing in the furnace.[2] Complete blending of the dry components prior to fritting is the main objective of mixing equipment. The surface treatment of the dusty powder is a secondary function. The process is one of apparent simplicity, but the physical properties of the separate ingredients of the batch dominate the solid–solid mixing process.

4.6 CHARACTERISTICS OF RAW MATERIALS FOR FRIT BATCH

In the raw materials commonly used in most fritting departments there can be wide differences in particle size distribution, density, particle shape and surface characteristics (arising, for example, from the residue of a mineral beneficiation processes) which can make homogeneous blending difficult. The completed mix will include all the elements in the formulation in the correct ratio. Insufficient mixing time and the loading of the mixer beyond its rated capacity will adversely affect the quality of the mixed batch. Poorly blended mixtures or the subsequent segregation of the batch will adversely affect the quality of the melted frit.

Particle size distribution. In mixing, the most important factor is particle size. When assessing particle size what is achieved is an approximation as though the particles possesed some definite geometrical shape, but chemical compounds and mineral grains, produced by crushing and grinding, have a wide variety of shape and size. If two materials differ widely in particle size, segregation during mixing is likely.

Bulk density. This is the weight per unit volume of the powdered material, expressed in terms of kilograms per litre. It is not a constant, but can be modulated by the degree of aeration and by the degree of mechanical packing induced by vibration.

True density or specific gravity. It is not the difference in specific gravity between materials which causes differences in the movement of particles in a mixer, but differences in the particle weights (and, therefore, the level of kinetic energy the particle can reach). A small sized particle with a high specific gravity can move, during mixing, more than a larger particle having a lower specific gravity.

Particle shape. Mixing plant can handle grains of every geometrical shape. Frit materials either have irregularly fractured outlines, are platey, agglomerated or crystalline and each shape has its characteristic behaviour in a mixer.

Surface characteristics. These are related to surface area. Fine materials might readily agglomerate, other ingredients can hold electrostatic charges and some particles might have an unduly rough surface. Any of these characteristics can inhibit smooth mixing. The intermixing of hygroscopic materials will be altered by ambient conditions and by the addition of moisture.

Flow characteristics. Both the angle of repose and the flowability[3] of a powder are measurable characteristics. Powders with a high angle of repose have lower flowability. Flow is also affected by moisture content and the distribution of coarse and fine grains.

Friability. During the handling of friable material, such as sodium carbonate, there is a tendency for particles to break into smaller grains which upsets the balanced spectrum of size relationships.

State of agglomeration. Some raw materials do not exist as independent particles but, by adherence, form clusters. These clusters might be only loosely formed and the degree of energy generated by the mixer within the batch will affect the extent of agglomerate breakdown and subsequent particle dispersion.

Moisture or liquid content. Although the raw material might have the appearance of a dry solid a small quantity of moisture might still be present. This can be advantageous, aiding dust suppression and reducing the risk of segregation (see Section 4.8). Red lead oxide is available with a carefully chosen quantity of moisture or an oil added to suppress dust while mixing and when charging the furnace yet avoiding interference with the homogenising process. Unfortunately the percentage of moisture in some raw materials is variable and constant assessment should be made so that compensation can be made to the recipe.

4.7 HOMOGENEITY OF BATCH MIXING

The theoretical end result of mixing will not be an arrangement where a grain of one component is always next to a grain of a different type.[4] Ideally for rapid melting of homogeneous frit, every grain should be adjacent to one of another type, but the end result of random tumbling will be a random mixture and this will be adequate in most fritting plants. The efficiency and costs of the whole frit-making operation can be related to the starting solids mixture. Various types of analysis, qualitative or quantitative, can be made to check the uniformity and completeness of mixing in the dry batch.[5, 6] Smearing a small sample with a spatula might be all that is necessary to determine whether the product will be satisfactory. Whatever evaluation test is chosen the sampling procedure is critical. The method of removing representative samples from the bulk mixing must be considered, and also the location, the number and the size (expressed as a fraction of the mass) of the samples.[6]

4.8 DEMIXING OR SEGREGATION

During mixing. Raw materials differing widely in physical properties are not easy to mix homogeneously in the dry state. Extreme differences in specific gravity, particle shape or grain size will result in natural segregation during mixing. Because of differences in their kinetic energy, heavier and smaller particles or round smooth grains tend to sink through lighter, larger or ragged grains. Particles with the greatest energy will move furthest and always to the edges of a poured pile. Since all dry mixing will be carried out under dust collection equipment loss of batch material as dust through excessive negative pressure is also a means of segregation. Agglomerated batch material should, therefore, be broken down before mixing to avoid demixing.

During transfer. Subsequent processing of the batch after mixing must be considered as part of the homogenising process and demixing in transit must be avoided. The efficiency of this process can only be measured after the point at which the batch is delivered to the furnace. Unfortunately, transfer systems usually involve vibration, and segregation takes place unless precautions are taken. With any vibration the larger, denser, grains tend to rise to the top of the batch—in effect the small grains fall between the larger grains and areas of different composition will develop where there are concentrations of one size of particle. A mixed batch discharged into a hopper will become segregated with coarse particles collecting around the perimeter. This becomes important on further discharge to the furnace because the finer material in the centre of the pile will leave the bottom of the hopper first. The transfer system should, therefore, have possible segregation points such as long drops, lengthy belt traverses, vibratory feeders or transfer stations reduced to a minimum. On this basis the mixer will be adjacent to the furnace or part of it.

Prevention of demixing. The addition of water alone to the mixing batch has a marked effect in reducing segregation, though the optimum amount added to the batch whilst avoiding caking needs to be determined.[7] Nevertheless, agglomeration can be encouraged to advantage. High alkali batches such as are required for low melting point borax frits respond to water additions differently from alkali-free lead silicate frit batch.[8] In weight terms the added amount of water will range up to 5%. This is added as fine spray when mixing is almost complete. The mixing plant is then operated until the batch is uniform.

4.9 FRIT MELTING

After blending, the batch is fed to the melting furnace, of which there are several types commonly in use.

4.9.1 Crucible Melts

On a very small scale, frits can be made by firing the raw materials in a crucible or even a thick biscuit pot (though an unsafe practice) in imitation of the old method where the batch was melted or "run down' in saggars in the pottery kilns. Larger sized laboratory melts can be made in specially built crucible furnaces but the operation gives quantities suitable only for development experiments. Once any degree of production is considered, frit made in bulk will be necessary and the complexity of the required plant means that the capital requirement for frit manufacture is large.

4.9.2 Bulk Melts

Early frit kilns were refractory-lined baths directly heated by coal, the flames passing from the firebox into the melting chamber. Heat reverberated down into the powdered batch until melting was complete. The modern direct equivalent is a continuous kiln which is suitable for long runs of a single composition. When fritting continuously the premixed ingredients are fed to one end of the box-shaped furnace heated with either gas or oil. The heat input is balanced so that molten frit flows from the kiln at the same rate as the raw batch is introduced.

Where large numbers of frits of different composition are required in relatively small quantities (perhaps 1–10 tonnes per run), a number of batch kilns offers the increased production flexibility that is required. Rotary furnaces are a common form of batch kiln used in frit manufacture. Such kilns consist of a cylindrical steel cylinder laid horizontally, lined with refractory, and geared to rotate as required. The batch is fed to the furnace through a centrally placed hole, subsequently sealed during the melting cycle. They are usually heated by the combustion of natural gas in the U.K.

In these and other designs of fritting furnace, combustion of oil or LPG or electric heating can also be employed as the means of heat generation.

Whatever the design of the frit kiln, the outer shell is lined with refractory bricks whose composition is selected to withstand the corrosive effects of the molten frit. The refractories employed invariably have a high silica content so that contamination arising from corrosion and erosion does not significantly alter the composition of the final frit. Continued exposure to the movement of molten frit does result in erosion of even the best refractories so from time to time they have to be replaced. The life of refractories depends upon the type of kiln, the composition of the refractories employed and the chemical nature of the frit itself.

For gas-fired kilns, a recent innovation has been the introduction of oxygen enrichment of the gas/air mixture. With this technique savings are made on fuel costs because the level of diluent nitrogen gas is reduced and as a result higher temperatures can be attained than would be the case with a simple gas/air mixture.

Depending upon the characteristics of the frit composition and the nature and operating conditions of the kiln employed, residence time of the batch in the kiln is in the range 1–20 hours. In gas-fired rotary kilns a residence time of 4 hours for leadless glaze frits is typical. Kiln operating temperatures are usually in the range 1250–1500°C for glaze frit manufacture.

There is a choice of furnace type for bulk frit melting. The kilns operate either intermittently or continuously and are heated by gas, oil or electricity.

Intermittent kilns. Intermittent furnaces receive the powdered batch ingredients directly from the hopper as a single charge. During this phase the heating source is idling to reduce turbulence as the powder falls into the hot furnace. The temperature at this stage is about 900°C, but this is quickly raised to full melting temperature of about 1350°C. Later, when a sample of molten frit withdrawn from the furnace shows that the glass-forming reactions are complete, the contents of the kiln are quenched by running a stream of the molten glassy frit either into a bath of water (the "bosh"), causing granulation, or between water-cooled steel rolls to form flakes. Alternatively quenching can be achieved by subjecting the molten frit to a combined blast of air and water mist. These granulation processes produce an easily ground form of frit.

The furnace temperature drops to 900°C and the cycle is repeated with a fresh charge of the batch.

It is usual to produce leadless transparent, leadless opaque and lead frits in separate furnaces, each dedicated to the manufacture of a separate frit type to avoid contamination. Cross-contamination is also minimised by arranging for frits of similar composition to follow each other in the production programme. When this cannot be arranged it is normal to consider the first charge of a new sequence as a "cleaner". After fritting, this cleaning charge is not used in glaze manufacture but is reserved to perform a similar cleaning service.

Continuous kilns. Continuously operated kilns have a constant supply of powdered batch ingredients fed into a melting chamber held at a temperature of about, say, 1350°C. The feed rate of powder into the kiln is matched to the rate at which the fully reacted glassy frit streams into the quenching machinery. At no time does the temperature in the kiln fall as in the batch process.

4.10 FRIT REACTIONS

Three distinct events occur during heating to bring about glass formation in the frit ingredients:

1. Chemical decomposition.
2. Diffusion of elements one with another.
3. Solution.

Heating is continued for a sufficient length of time to ensure a homogeneous product free from unmelted inclusions. For some purposes complete fusion is unnecessary; all that is needed is for the requisite chemical combinations to have occurred whereby the water-soluble substances are made wholly insoluble. The liquid, molten, frit is then quenched.

The complete frit, if properly balanced in its acids and bases, is virtually insoluble in water and there are other properties dependent upon this balance. Glasses high in alkalis and low in acidic components have low resistance to chemical attack. It is the relative proportion of these ingredients in the frit which determines the working characteristics of the glaze of which it is a part. However, frits do not have uniquely fixed compositions and thousands of frits each with a different composition are produced by the ceramics industry. In theory the number of possible frits is limitless, but ceramic practices exercise constraint on their variety and number. The essential requirement for easy melting in bulk, wide temperature range in the glost fire, good chemical resistance or durability and the use of available raw materials at an economic price, often lead to roughly equivalent recipes and consequently some compositions are found to be "standard" in certain areas of glaze technology.

There are four main types of frit:

1. Transparent borosilicate (leadless).
2. Opaque borosilicate (leadless).
3. Lead bisilicate.
4. Lead borosilicate.

4.11 MELTING OF FRIT

The relative importance of reactions which occur in the frit furnace are characteristic of the type of frit and the constitution of the composition being melted. The proposed end use of the desired glaze will define the quality of the raw materials chosen for the frit recipe.

In the role of glass network-former most frits contain silica, many contain boric oxide and some specialised frits contain oxides of phosphorus. These network-formers are made to react by heat with network-modifying oxides, both fluxes and stabilisers, the precise proportions of which determine the properties of the final melted frit. Many aspects of the thermal behaviour of chemicals, have been examined either singly or in mixtures as they melt to form glassy materials,[9,10] and there is published work which enables new developments in glaze frit composition to be transferred from laboratory scale to commercial manufacture.[11,12] Discussion as to the choice made between different raw materials providing particular elements appears in Section 3.2.

The stages in the formation of a vitreous frit can be followed by differential thermal and gravimetric analysis. As the temperature of the batch ingredients in the furnace increases, moisture in the raw materials is evaporated. With further heat input the materials begin to decompose, liberating water of crystallisation, chemically bound water, F_2, CO_2, O_2, SO_2 and SO_3. Some ingredients, for example zinc oxide and lead oxide,[13] begin to interact in the solid state before true melting begins. Evidence for early endothermic or exothermic reactions in the batch can be found by differential thermal analysis.[14,15] Most batch reactions are endothermic. The continued absorption of heat by the powdery ingredients can result in liquid eutectic mixtures,[16,17] between some components, so initiating more widespread melting. Although the melting point of sodium carbonate is 850°C, during the formation of lead free frits, a eutectic mixture forms between sodium carbonate and calcium carbonate which melts at 775°C. Sodium borate melts at 742°C.

As decomposition proceeds, reaction gases are evolved rapidly and this high rate of evolution combined with the great volume of gas agitates the incipient molten mass, encouraging further glass-forming reactions and clearance of bubbles. As the temperature rises to about 1000°C the melting reactions proceed rapidly with the remaining, slowly soluble, refractory raw materials. The melting process consequently becomes controlled by the rate at which the remaining refractory grains (for example silica or zircon) react and dissolve[18] and by the rate at which other liquid phases react. Throughout these latter stages, as the rate of temperature increase slows and the viscosity of the heterogeneous melt changes, release of gas from the viscous vitreous mass must continue. Continued heat input beyond this point, though difficult, is necessary to complete the homogenisation and degassing of the molten frit.

When all the crystalline materials have been incorporated into the newly-formed frit there remains a large amount of gas distributed throughout the melt in a myriad of bubbles. These bubbles escape throughout the frit melting sequence. For some glaze compositions it is not a prerequisite that all entrapped gas bubbles should be removed during frit melting (some bubbles are eliminated from frit during the glaze grinding process and some gas will escape during the glost fire), but to avoid trouble later as many bubbles as possible must be removed in the frit furnace. The time taken to complete this refinement is variable.

Stoke's Law relates the rate of rise of a bubble to the surface of a liquid to several factors:

$$v = \frac{2}{9} \, gr^2 \, \frac{\rho}{\mu}$$

where v = velocity of rise,
g = constant gravitational,
r = radius of bubble,
ρ = density of medium (i.e. frit),
μ = viscosity of medium (i.e. frit).

Such a relationship defines the optimum conditions for fritting and implies that larger bubbles will escape most readily when the molten frit is least viscous. Low viscosity in the molten frit can be achieved by raising the furnace temperature to a permitted maximum, and some lowering of the density is achieved at the same time. At high temperatures the solubility of gas in glass is low and diffusion rates are high. Other processes[19-21] can be invoked to encourage bubble removal from massive frit melts without invoking bubble flotation and escape.

Surface tension is an important factor in the removal of bubbles:

$$P = \frac{2\sigma}{r}$$

where P = excess pressure
σ = surface tension,
r = radius of bubble.

Small bubbles dissolve when the diameter is low enough for the surface tension to increase the internal pressure.[22]

The efficient melting of frit batch is affected by a number of factors.

Furnace temperature. For commercial frits the time to achieve freedom from undissolved material decreases with increase in temperature.[23] Compensation for a reduction in melting temperature can be made by increasing the residence time of the frit in the furnace.

Particle size of frit components. Assuming we have a fixed melting temperature, then for any particular frit composition, melting rate can be altered by changing the grain size of the components.[24] The smaller the grain size of the constituents the faster will be the frit-forming reactions.[25]

There are limits to this general rule. Ultrafine materials such as the clay minerals, tin oxide, or titania have very small apparent particle diameters, but they readily agglomerate into loosely bound granules. The resulting porous aggregates resist the transmission of heat and being less dense, float on the surface of the melts, resisting solution. Carbon dioxide, evolved from the glass-forming reactions, escapes through channels around remaining grains of the batch but, as glassy liquids form, these paths are sealed. Finely ground materials can form only narrow pathways which become sealed more rapdily, suppressing the decomposition reaction between silica and sodium carbonate.[26] Alternatively if the flux components in the frit recipe are very fine, refractory grains are coated evenly with a potentially active layer, so encouraging the onset of early reactions.

Frit composition. The solubility in the molten frit of the principal gases evolved during the glass formation process, carbon dioxide, water vapour and sulphur dioxide, depends upon the composition of the frit. As the alkali content of the frit increases so does the solubility of water vapour in it. It is not, therefore, a safe, easy alternative when trying to free a frit from bubbles to increase the alkali content (reducing the viscosity of the melt chemically) rather than to increase the melting temperature (reducing the viscosity of the melt thermally). Oxygen is very soluble in glass.[27]

Degree of demixing. The ease of melting is directly related to the degree of segregation in the batch which takes place between mixing and furnace charging mechanisms. Although in frit melters, stirring (and, therefore, remixing) is absent, most furnaces can cope with minor segregation and produce homogeneous frit. Demixing on a large scale, leaving some refractory ingredients undissolved to float as a scum on the surface of the melt, will prevent the formation of a glass frit containing all the component oxides.

4.12 FRIT QUALITY TESTING

Before quenching. When a frit composition has been melted for some time the homogeneity of the product is checked by withdrawing a sample of molten frit from the kiln and inspecting a drawn thread when cool for its clarity and for its freedom from inclusions of undissolved materials. With a continuous frit kiln, the conditions of operation, e.g. feed rate, operating temperature, need to be manipulated until the required quality is attained.

Before glaze preparation. It is essential that the quality of the frit is first determined because when employed in glazes, frits are generally a major raw material in the glaze mill batch. The complexity of the testing required will depend on the final application in which the frit is employed. For some technical ceramic applications (e.g. in electronics) full chemical analysis on each batch may be required, but this would not normally be necessary when the final product is tableware glaze. A comparison of fusion and flow characteristics of each new batch of frit against an accepted standard is, however, a test which would be applied to all products. Other tests which may be applied are:

Thermal expansion (including *Tg* and *Mg* determinations),
Solubility (particularly lead solubility),
Specific gravity,
Colour (when applied and fired as a glaze; this test would particularly be applied to opaque frits).

REFERENCES

1. Harwood, C. F. *Glass Industry,* **56** pp. 12–15, (1975).
2. Ed. Doyle, P. J., *Glass Making Today* pp. 152–155. Portcullis Press (1979).
3. ASTM Test B213–48, Flow rate of metal powders.
4. (Eds.) Drew and Hoopes, *Advances in Chemical Engineering,* Vol. II. pp. 259–265. Chapter on "Mixing of Solids" by S. S. Weidenbaum, pp. 211–321. Academic Press 1958.
5. Harwood, C. F. *Glass Industry,* **56** pp. 12–15, (1975).
6. A.I.Chem.E. Testing Procedure, "Dry solids mixing equipment" (1979). American Institute of Chemical Engineers, 345 East 47 Street, New York 10017.
7. Parkin, M. and Turner, W. E. S. "The influence of moisture on the mixing of soda–lime–silica glasses". *J. Soc. Glass Technol.,* **10,** 114–129 (1926).
8. Parkin, M. and Turner, W. E. S. "The influence of moisture on the mixing of batches for potash–lead oxide–silica glass". *J. Soc. Glass Technol.,* **10,** 213–220 (1926).
9. Kröger, C. *et al., Glastechnische Berichte* **15,** 335–346, 403–416, (1937); **22,** 86–93, 248–261, 331–338, (1948/9); **25,** 307–310, (1952); **26,** 346–353, (1953); **27,** 199–212; (1954); **28,** 51–57, 89–98, 426–437, (1955); **29,** 275–289, (1956); **30,** 222–229, (1957).
10. Morey, G. W. and Bowen, N. L., *J. Soc. Glass Technol.,* **9,** 226–264T (1925).
11. Lyle, A. K. *J. Amer. Ceram. Soc.,* **28,** 282–287 (1945).
12. Cable, M., Rasul, C. G. A. and Savage, J. *Glass Technology,* **9,** 101–104 (1968).
13. Preston, E. and Turner, W. E. S. *J. Soc. Glass Technol.,* **25,** 136–149 (1941).
14. Clark-Monks, C. *Glass Technology,* **13,** 138–140 (1972).
15. Warburton, R. S. and Wilburn, F. W. *Physics Chem. Glasses,* **4,** 91–98 (1963).
16. Morey, G. W. and Bowen, N. L. *J. Soc. Glass Technol.,* **9,** 226–264T (1925).
17. *Phase Diagrams for Ceramists,* 1964, 1969, 1975 and 1981. American Ceramics Society.
18. Cable, M. *Glasteknisk Tidskrift,* **29,** 11–20 (1974).
19. Cable, M. *Proc. 8th Int. Cong. on Glass,* London, pp. 163–168 (1968).
20. Cable, M., Rasul, C. G. and Savage, J. *Glass Technology,* **9,** 101–104 (1968).
21. Ed. Doyle, P. J., *Glass Making Today,* pp. 170–175. Portcullis Press (1979).
22. Ed. Doyle, P. J., *Glass Making Today,* p. 172. Portcullis Press (1979).
23. Preston, E. and Turner, W. E. S. *J. Soc. Glass Technol.,* **24,** 124–138T (1940).
24. Boffé, M. and Letocart, G. *Glass Technology,* **3,** 117–123 (1962).
25. Potts, J. C., Brookover, G. and Burch, O. G. *J. Amer. Ceram. Soc.,* **27,** 225–231 (1944).
26. (Ed. Doyle, P. J.), *Glass Making Today,* p. 166. Portcullis Press (1979).
27. Morey, G. W. and Bowen, N. L. *J. Soc. Glass Technol.,* **9,** 226–264T (1925).

Chapter 5
The Glaze Manufacturing Process

5.1 GRINDING

Manufacture of glaze involves the production of a finely ground and intimate mixture of raw materials such that each portion of the batch which might be sampled exhibits the same properties. Although there are means of applying glaze in the dry or even molten state, it is usual within most sectors of the ceramic and other industries to employ a dispersion of the ground raw materials suspended in water. This glaze slip might consist, in addition to the carrier medium, water, solely of one component (a ground frit) or a mixture of many materials including frits, ceramic minerals and insoluble manufactured chemicals as well as some very small quantities of minor materials, such as flocculants, deflocculants, hardeners and organic dyes. Other carrier media, e.g. alcohol, might be employed in specialist applications.

The most widely employed glaze manufacturing methods involve (i) mixing the ingredients, (ii) particle size reduction, (iii) dispersion in water, (iv) removal of unwanted materials and (v) the addition of minor quantities of chemicals to modify certain physical properties of the glaze slip. The first three processes are carried out together during wet grinding.

If glazes prepared by wet grinding are subsequently dried and supplied by a commercial glaze miller to his customer in powder form, a further dispersion process will have to be performed by the user to obtain a glaze slip ready for use. Alternatively, a glaze manufacturer can supply a weighed-out charge of glaze which has been premixed, without comminution, in the dry state. Wet grinding in, for example, a ball mill would then be carried out by the user to effect the necessary particle size reduction and dispersion in water. Dry dispersion of finely pre-ground raw materials or dry grinding are other techniques which can be employed with the resultant product being later dispersed in water if conventional glaze application techniques are to be employed.

Grinding in terms of glaze preparation is the pulverisation of the raw materials by impact and abrasion. There are several classifications of grinding related to the particle size of the final product. Glazes are classed as finely ground powders because they have a preponderance of particles less than 75 μm (or less than 200 mesh). The mineral components of a glaze recipe are products of intermediate grinding, being less than 0.5 mm (or less than 28 mesh).

The machinery most frequently used in glaze production is a "tumbling" mill comprising a rotating cylinder partly filled with freely moving impact-resistant shapes. Equipment of this type is called a ball or pebble mill. As the cylinder rotates, the particles to be ground are struck hard blows and are abraded against each other and against nearby hard surfaces.

An alternative method of energising the mill is to vibrate it rather than rotate it, and in using this principle a range of particle size reduction plant for ceramics, called vibro-energy mills, was developed.

5.2 BALL MILLING

Wet ball milling is the most commonly used industrial method for producing glazes. Ball mills are closed cylindrical vessels (apart from a lockable loading/discharge plug) which are rotated about their horizontal central axis. They are rugged in construction and not difficult to operate. Sizes suitable for laboratory trials and bulk production grinding are available. Production mills are usually constructed in metal and are lined internally with an abrasion-resistant material, the nature of this lining being influenced by the product being ground. In order to bring about size reduction of the glaze ingredients, ball mills contain "media" in the form usually of spheres, but sometimes in the form of cylindrical rods or similar shapes. The combination of mill lining and grinding media is such that the action of one on the other, or of the media shapes on each other, does not result in significant contamination of the product being ground.

Particle size reduction arises from the cascading action of the media on the grains or lumps of glaze batch as the mill rotates. Ball milling, even when conditions are optimum, is a relatively energy-inefficient process and, therefore, it is essential to ensure that various factors which can affect grinding efficiency are controlled. Factors which affect efficiency, contamination and speed of grinding include:[1-3]

1. Type of mill lining.
2. Type and size of grinding media.
3. Weight and ratio of each of grinding media, glaze components and water in the mill.
4. Mill speed.
5. Particle size of feed material.
6. Consistency and "hardness" of the feed ingredients.

5.2.1 Mill Lining

The contents of the mill will tend to wear the lining leading to contamination of the batch. Therefore, it is necessary to select a lining material which has good resistance to abrasion by the grinding media and glaze components and which resists impact by the tumbling media. Some contamination is unavoidable and the use of silex (a natural, high silica rock), porcelain or alumina as lining materials is long established. Contamination of the mill contents with fragments of these materials is generally not serious because their oxide components are themselves present in most glaze compositions. Rubber linings[4] have become increasingly used in recent years because, in combination with high-density media, they can lead to marked reductions in the time needed to reduce the particle size to the required degree of fineness for a particular glaze. The linings in use have good wear resistance and reduce the level of noise during operation of the mill. Contamination of glaze by fragments of rubber is not generally a problem since this organic material can burn away from transparent glazes during the glost fire. Unfortunately, when present in some coloured glazes, such points of reducing condition localised around the grains of decomposing rubber can affect the pigmentary power of the glaze stain. In chrome–tin crimson glazes such conditions would lead to "white spot".

5.2.2 Type and Size of Grinding Media[5]

Factors which influence the choice of grinding media employed in ball mills are toughness, impact resistance, abrasion resistance, density, shape, homogeneity and freedom from voids. In wet ball milling, the density of the media relative to the density of the glaze slip is an important factor affecting the speed of particle size reduction: the greater the density of the media compared with that of the slip, the faster will the milling proceed for slips of similar viscosity. Glaze slips of optimum viscosity for grinding typically have densities of 1.6–2.5 g ml^{-1} depending on the specific gravities of the component raw materials. A leadless glaze could for example be readily ground at 1.6 g ml^{-1} but for a high lead glaze (65% PbO content) a slip density of approximately 2.5 g ml^{-1} would be necessary to ensure efficient grinding, although its slip viscosity would be similar to that of the leadless glaze cited. The specific gravities of media commonly employed for the grinding of glazes are as follows:

Flint pebbles	2.5–2.6
Porcelain	2.3–2.6
Steatite	2.6–2.8
High-density alumina	3.3–3.6

Clearly, the use of flint pebbles for the grinding of the high lead glaze previously mentioned would result in little particle size reduction of the glaze components. The individual media will have a tendency to suspend in the slip due to the similarity in the specific gravities of media and slip; consequently little grinding action will occur because the necessary cascading action and impacting of the media will be absent. Floating balls or pebbles do not grind.

Alumina media having a higher density than the other ceramic media listed are widely employed because they produce more rapid grinding. They are, however, relatively expensive and flint pebbles, although having a lower density are still used, being cheap and having good hardness characteristics.

For wet ball milling, spheres are the preferred media shape, the pebble shape of flint media resulting in relatively inefficient grinding.

The selection of grinding media should take account of the mill lining available. Certain combinations of lining and media can result in excessive contamination. Combinations typically employed for the grinding of glazes are rubber lining with high-density alumina and silex lining with flint pebbles.

The relative sizes of the grinding media have a bearing on the rate of particle size reduction. It is normal practice to use balls of three sizes, with the largest having a diameter no more than three times the diameter of the smallest media, otherwise impact damage on the smallest media will be a serious problem and result in unacceptable levels of contamination of the glaze with refractory particles. The intermediate sized media generally account for at least 50% by weight of the total media charge, the remainder being made up from equal weights of the other two sizes.

Large balls have a relatively greater effect on the initial grinding of a glaze when there is a preponderance of large grains of frit or mineral. Once particle size reduction has proceeded to a certain point, effective further grinding is brought about through the action of the smaller sized media. Compared with the smaller sized balls, the larger media are greater contributors to wear and temperature increase in the mill.

The average size of the media should, for optimum grinding efficiency, be related to the size of the glaze ingredients. A larger feed size requires a larger average media size, although the nature of the individual materials being ground should also be considered. For grinding glazes in batches of up to 1 tonne, the media are not normally of a size greater than 50–60 mm diameter. When high-density alumina media are used, somewhat smaller-size balls will suffice due to their higher inertia and more efficient grinding action.

5.2.3 Mill Loading

In wet ball milling, the loading of the mill with glaze charge, grinding media and water is critical and the quantities of each must be closely controlled if optimum grinding efficiency is to be attained. Experience has shown that for best results the grinding media and intervening voids should occupy approximately

55% of the mill volume. The media should occupy this volume no matter what size of mill or which type of medium is employed. Consequently, when high-density alumina media are employed, the media weight in the mill will be significantly greater than if flint pebbles were used.

The volume of the glaze batch feed materials should fall within the range 11–18% of the total mill volume. The actual volume of the batch for optimum efficiency is related to the nature of the grinding media; where high-density media are employed, the batch materials can occupy a greater proportion of the mill volume than is the case when, for example, silica pebbles are used. A greater volume of batch can be tolerated when high-density media are used because of the greater momentum which they possess during the grinding process, other factors being equal.

The amount of water charged to the ball mill has a significant effect on grinding efficiency and care must be taken to ensure that, for each glaze composition, the correct amount of water is used. For non-plastic materials, approximately equal quantities by volume of water and batch materials usually leads to good grinding efficiencies. When clays are present in the batch, a proportionately greater amount of water will be necessary to counteract the increased viscosity resulting from the presence of these plastic materials. Other types of suspending agents might be used and in these cases also an increase in the proportion of added water may be necessary. For wet ball milling, the water therefore usually occupies 50–55% of the total slip volume.

If too much water is present in the mill, the particles of the raw material tend to escape from the abrading surfaces of the media because they are washed away from the impact points by the water. A viscous film of slip cannot form around the media, resulting in additional wear of the latter and of the lining material.

When too little water is added to the mill, the highly viscous slip also impairs the particle size reduction process because the media impacts are smothered. Thus the quantity of added water can be used to control the grinding process through its effect on the path taken by the balls as they move over the internal surface of the mill lining.

5.2.4 Specific Gravity of Glaze

The importance of the specific gravity of the glaze being ground is illustrated in the following table where suitable ratios of water and mill materials are given for glazes of specific gravity 2.5 and 4.5. These specific gravities could represent typical leadless and very high lead glazes respectively.

S.G. of dry glaze	Wt. % water	Wt. % solids
2.5	32.8	67.2
4.5	21.4	78.6

Apart from completely fritted glazes, two or more materials will constitute a glaze formulation and, therefore, a glaze can contain a number of materials each with different specific gravities. However, during the grinding process separation of the components is not a problem and the mean specific gravity of the composition can, therefore, be used when calculating the required amounts of water and glaze batch.

5.2.5 Mill Speed

As a ball mill rotates, the contents, under the influence of gravitational, centrifugal and frictional forces, move, to some extent, relative to the mill lining. This movement depends upon the speed of rotation (Fig. 5.1).

Fig. 5.1. Movement of tumbling balls in the layers adjacent to a mill lining.

At very low rotational speeds, the media and the glaze components will remain at the bottom of the mill because gravitational forces predominate. Little if any grinding of the glaze would result, but wear of the lining would occur. At the other extreme, if the mill speed is too fast, centrifugal forces drive the grinding media and glaze charge outwards onto the mill lining where they remain stationary relative to the lining. Because there is no relative motion between the media and charge, there is neither impact nor abrasion and no particle size reduction occurs. The minimum mill speed, at which centrifugal forces cause the media to remain continuously in contact with the lining, is known as the critical speed. Clearly a situation between these two extremes is required for efficient grinding and in practice a mill speed which is approximately 60% of the critical speed has been found to give best results. At such a speed, the glaze materials and the media are carried upwards by frictional forces in the mill to the correct height before cascading down onto the underlying material. Particle size reduction then occurs by attrition as the media, coated with glaze material, tumble over each other (Fig. 5.2).

At a slightly higher rotational speed the charge will be carried too high on the circumference of the mill before collapsing. This inefficient condition is called cataracting (Fig. 5.3) and leads to high rates of media wear and to high levels of product contamination. The ideal angle, between the horizontal through the axis and the point where the outer balls leave the mill lining, is 45°.

The linear speed of the mill casing for a given speed of rotation is greater for those mills with a larger diameter. Consequently large diameter mills run at

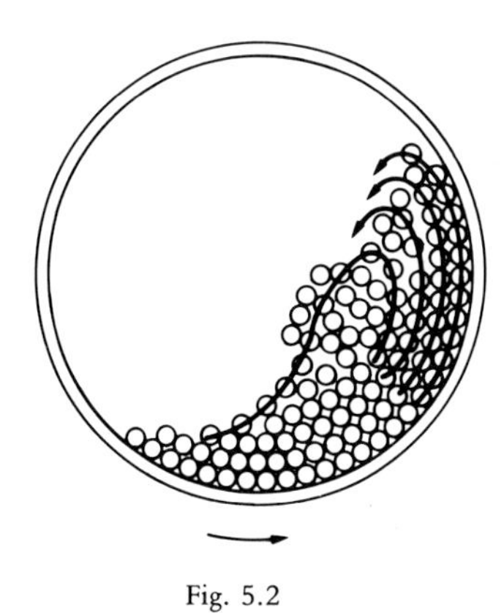

Fig. 5.2

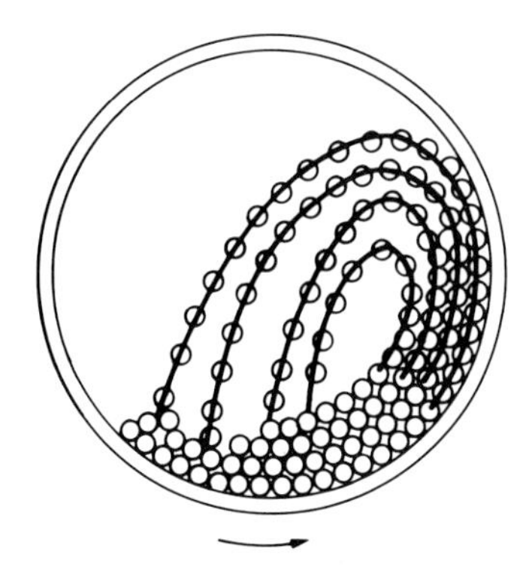

Fig. 5.3

lower rotational speeds than do small mills. Although wear of the lining and media can be alleviated, it cannot be eliminated; therefore, in addition to ensuring that conditions are such that wear is minimised, action is taken to counter the effects of wear. With continuing wear in the mill lining, the internal volume of the mill increases and if no action is taken to modify the weights of the media and glaze charge, non-optimum conditions will result. An increase in mill volume of 10–20% can readily occur as a result of the wear of a mill lining and the mill will then be operating in an undercharged mode.

Wear of the grinding media naturally results in a reduction of their size and ultimately in the presence of very small media, capable of being broken up into fragments by the larger media in the mill. It is, therefore, necessary to remove periodically such small media to avoid their fragmentation. The required media volume is maintained by adding a balancing quantity of the large-sized media. Media of less than 20 mm diameter should be removed from ball mills with nominal internal volumes of more than 1200 litres.

Periodic (2–3 month) inspection of mill lining and grinding media are, therefore, necessary to ensure that optimum conditions are maintained. Control of the charging weight of "balls" in the mill can be done by observing their level in the mill.

5.2.6 Dry Grinding

Glaze is manufactured to a limited degree using a dry grinding technique. This is advantageous with glazes which are either to be transported large distances rather than used at the site of manufacture or which are to be applied dry by, for example, flame spraying or fluidised bed methods. In addition, for glazes which exhibit an appreciable solubility in water, production by dry grinding techniques eliminates problems arising from the presence of soluble species in a water slip.

Efficient wet grinding, resulting as it does in an appreciable rise in the temperature of the slip, would tend to exacerbate problems with a formulation exhibiting a marked solubility. The solubilised salts are usually alkaline and these markedly affect the rheology of glaze systems containing clay.

Much of what has been previously stated concerning wet milling processes for glaze manufacture applies equally to dry grinding operations, but there are slight differences in technique to ensure optimum size reduction efficiency. For example, in dry ball milling the media and intervening interstices should occupy 45–50% of the mill volume compared with a figure of 55% for wet grinding.

5.3 VIBRATION GRINDING[6, 7]

Vibration mills can be employed both for wet and dry grinding and have found limited application in the production of glazes (e.g. sanitaryware glazes). Both horizontal and vertical mills are available. High-frequency vibration forces are produced in the mill by the use of out-of-balance weights attached to a horizontal or vertical shaft, driven by an electric motor.

The Podmore–Boulton Vibro-Energy mill has a vertical grinding chamber which is vibrated three-dimensionally by the application of a circular displacement in a horizontal plane with a superimposed vertical oscillation. With this arrangement, it is possible to attain close packing of the special grinding media which are preferably cylindrical in shape. This type of mill is fully charged with media, which themselves occupy 75% of the chamber volume, the remaining 25% (i.e. the voids) being occupied by the material to be ground. Close packing of the cylindrical media means that coarser material is ground preferentially, so giving a product with a narrower particle size distribution than would have been achieved if conventional ball-milling techniques had been employed. In addition, vibration grinding results in substantially reduced milling times compared with ball milling. In a vertical mill such as the Vibro-Energy mill, mixing is not efficient, the energy being utilised primarily in particle size reduction. For this reason, when using such a mill for wet grinding, some premixing of the ingredients or subsequent recirculation of the slurry should be carried out to ensure the final ground product is adequately mixed.

The standard mill lining for a Vibro-Energy mill is abrasion-resistant rubber and the grinding media used are usually alumina or sintered zircon. The surfaces of these should be polished. To allow adequate movement of the media during wet-grinding with the Vibro-Energy mill, the slurry should have a viscosity of approximately 200 c.p.s. For optimum grinding, the feed size of the glaze components should be less than 250 μm (60 mesh).

In the horizontal vibration mill, the media and intervening voids occupy 85–90% of the mill volume with the charge occupying 100–120% of the voids volume for optimum efficiency. For this type of mill, circular motion of the media (i.e. equal horizontal and vertical amplitudes) gives optimum results and spherical media are therefore preferable. The number of size reducing impacts is greater than that achieved in ball milling, although the impacts are usually less intense. However, the outcome is that the horizontal mill gives speedier grinding. In contrast to the vertical mill, the horizontal mill mixes the batch ingredients very efficiently.

Continuous operation of vibration–grinding mills is feasible. In continuous wet grinding with the Vibro-Energy mill, product control can be exercised by the use of metering pumps and in dry grinding an air-classifier can be utilised. Due to the relatively fast speed of size reduction, vibration grinding can lead to grain size control difficulties. As little as 10 minutes additional grinding in a Vibro-Energy mill can mean the difference between a product being in or out of specification. This problem can be alleviated by re-circulating a large batch of glaze between the mill and a stock tank which holds a much greater volume of slip than can the mill itself.

5.4 MILL-OPERATION

With the mills correctly set up and operating conditions established, manufacture of the glaze involves weighing out the ingredients which might include frit, ceramic minerals, manufactured chemicals, pigments and small quantities of additives, blending if necessary, charging the batch to the mill with, for wet grinding, the requisite amount of water and milling to produce the glaze at the required particle size specification.

In order to prevent cross-contamination, coloured and non-coloured (i.e. transparent or white opaque) glazes are normally ground on separate mills. Lead based glazes and leadless glazes are usually processed in separate mills for the same reason. If it is not possible, due to lack of available mills, to keep different types of glaze separate, then the thorough cleaning of the mill after each charge of glaze has been produced is essential. In the normal wet grinding processes, cleaning is either by washing the contents of the mill out with water, or if necessary by grinding a cleaning charge of an innocuous raw material which is unlikely to contaminate later mill charges.

In addition to the major glaze constituents, small amounts of other materials are often added at the mill. For transparent low solubility glazes intended for application to white bodies (e.g. bone china), a minor amount of blue cobalt aluminate stain is added to offset the yellow coloration imparted to the glaze by the lead frit employed. A similar stain addition is often added to white opaque sanitaryware glazes to produce a less yellow-toned white. In these cases, it is normal practice to add the stain as a predispersed slip to the glaze batch prior to wet grinding, otherwise pigment specks may subsequently be seen in the fired glaze. Whatever the type of grinding process, as new surfaces are generated by the comminution process, their activity tends to slow down further grinding. In order to counteract this, the addition of grinding aids to dry milling processes or of dispersants to wet milling operations can be advantageous. For wet grinding, additions of deflocculants can result in higher density slips. Small additions of binders to aid glazing and subsequent handling of ware may also be added at this stage, although break-up of certain binders occurs at high mill temperatures. If these binders are susceptible to degradation by bacteria, an addition of biocide will be necessary if the glaze is to be stored in the slip state for anytime.

For wet ball milling, raw materials are often graded less than 0.5 mm (28 mesh) to less than 150 μm (100 mesh) except for frits which in crystal form can reach 1 cm. All the materials in the recipe or formulation will be charged to the mill together, although for a high frit content recipe, the frit alone might be ground first before adding the remaining ingredients and taking the grinding process to completion. For some special glaze effects certain components might be added toward the end of the grinding process. These materials could be coarse grained and it will be desirable only to mix them uniformly into the slip without markedly altering their particle size characteristics.

Depending on the nature of the starting materials, the type of mill lining and grinding media employed and the desired particle size characteristics of the product, grinding times in wet ball milling operations range from 6 to 18 hours, the shorter milling times being achieved with high-density media. In Vibro-Energy mill operations with high density media, 1$\frac{1}{2}$ hours would be a typical grinding time required to achieve a similar degree of grinding, although coarse starting materials such as crystal frit must be preground to approximately 250 μm (60 mesh) before charging the mill. For ball milling, time is not the best indicator of the progress of the grinding process because slipping of the motor belts, by which mills are rotated, can occur. Counters which record the number of revolutions of the mill are a better means of control of the process at the factory floor level.

When Vibro-Energy mills are used, optimum vibration characteristics can be adjusted by altering the amount of media present in the mill rather than modifying the drive units. Consequently such mills are not initially charged such that they are brim full.

Particle size measurement is the means by which grinding processes are ultimately controlled and there are many methods available to determine different measures of particle size. These are discussed in section 10.1. For any new glaze formulation, particle size determinations will normally be made during the grinding of the first bulk charge so that simpler means of shop-floor process control can be formulated. Such control may be based on the number of cylinder revolutions (for ball milling) or time. Once these process control indicators have been obtained, subsequent batches of the same glaze can be produced by maintaining the same milling conditions or after making such modifications as are dictated by testing subsequent batches for particle size. Glazes are ground so that 60–90% of particles (by weight) are below 0.01mm diameter.

For any charge of glaze, particle size measurements on a sample of the milled product will be made to confirm that the material is within specification. If the glaze is found to be underground, additional milling would be necessary to bring the product within the specification. Except for glaze formulations which require coarse ingredients to be added at this stage, glaze slips found to be within their particle size specifications will be ready for removal from the mill. Such slips can be run from the mill by gravity or pumped into tubs or storage arks. To remove undesirable coarse fragments and any magnetic particles, it is normal practice to pass the slip through sieves and magnets prior to its passage to the storage containers. For tableware glazes, the sieves employed are generally in the range 125–75 μm (120–200 mesh), the finer meshes being employed for glazes subsequently to be applied more thinly to ceramic bodies (e.g bone china). If the slip in the mill is viscous, an appreciable proportion of it will be left within the mill. However, this can be largely removed by the addition of some water, taking care to ensure that an excessive amount is not added or this may result in the whole batch having a slip density less than that required in the subsequent glazing process. If too low a slip density does result, remedial action (e.g. letting the glaze stand for some days and then removing supernatent water) will be necessary. As glaze viscosities are lower at higher temperatures, it is advantageous to empty the mill while the contents still retain the heat generated during the milling process.

After running the bulk of the glaze from the mill, any remaining slip must be removed before a different glaze formulation is next processed on the mill. This is clearly most important with coloured glazes and when a leadless glaze is to follow a low-solubility lead glaze. Such cleaning can be achieved by adding water, and running the mill for a short time before removal of the watery slip. Two or three such washing operations may be necessary to achieve the required level of cleanliness.

After removal of the glaze batch from the mill it will be tested (after removal of water) against a standard reference sample of the same formulation to compare the fusion and flow characteristics. The samples should be fired to the temperature at which the glaze will subsequently be fired after application to ceramic ware. The cycle employed should be such as to show up differences which would be evident in the commercial firing. Such tests indicate whether the batch of glaze is too "soft" or too "hard" against the standard. Remedial action can be taken to bring any non-standard material within the required specification by suitable additions of fluxing (e.g. frit) or refractory materials (e.g. clays). The nature of any such additional materials will take into account the desired surface characteristics of the glaze (e.g. mattness or glossiness) as these properties could be affected if inappropriate modifying agents were added.

Depending on the use to which the glaze may be put, other tests (e.g. thermal expansion, colour measurement) may be performed and remedial action can be taken if necessary (e.g. colour modification).

When all important characteristics of the glaze batch have been shown to match those of the standard, it is normal practice to test the glaze by applying it to a ceramic substrate and firing the glaze under the production conditions or a controlled laboratory equivalent. Indeed if the ceramic manufacturer produces his glaze on site, this assessment of the glaze can, with advantage, be performed at an earlier stage.

For this latter test and subsequent use, the glaze will usually be adjusted to a specific slip density (by addition or removal of water) and standard viscosity (by the addition of flocculant or deflocculant). An addition of a cellulose-based hardner may be made to control the drying rate of glaze on the ware and impart hardness to the dry glaze coating subsequently produced. Prior to bulk use, an addition of water-soluble organic dyestuff might be made to distinguish the glaze from other glaze slips in use in the factory. Most white opaque and transparent glazes look very similar in the slop state and differentiation with organic dyes is necessary. At the very low levels of addition (approximately 0.005% on the slop weight), rheological properties are not affected and the organic material readily burns away during the glost fire.

REFERENCES

1. Laurs, A. *Interceram.* **19,** 31 (1970).
2. Slinn, G. and Rodgers, K. *Interceram.* **29,** 398 (1970).
3. Melandri, M. *Ceramica Informazione* **4,** 70–73 (1969).
4. Söderström, H. *Ceramic Powders* (Ed. P. Vincenzini), p. 283. Elsevier Science (1983).
5. Davies, H. E. *Bull. Amer. Ceram. Soc.,* **32,** 209–211 (1953).
6. Podmore, H. L. and Beasley, S. G. *Chem. and Ind.* pp. 1443–1450 (1967).
7. Beasley, E. S. G. and Slinn, G. G. *Ceramics* **18,** No 226 12–20 (1967).

Chapter 6

The Thermal Expansivity Correlation in Glaze and Body

6.1 GLASS TRANSITION RANGE

It is convenient when discussing glaze fit to assume that the glaze is a homogeneous glass.

Consider the behaviour of a molten glass cooling from the liquid state with random molecular groupings. The diagram (Fig. 6.1 similar to Fig. 1.1 on page 3) shows the relationship between volume and temperature as the molten material cools from the liquid state to the glassy state.

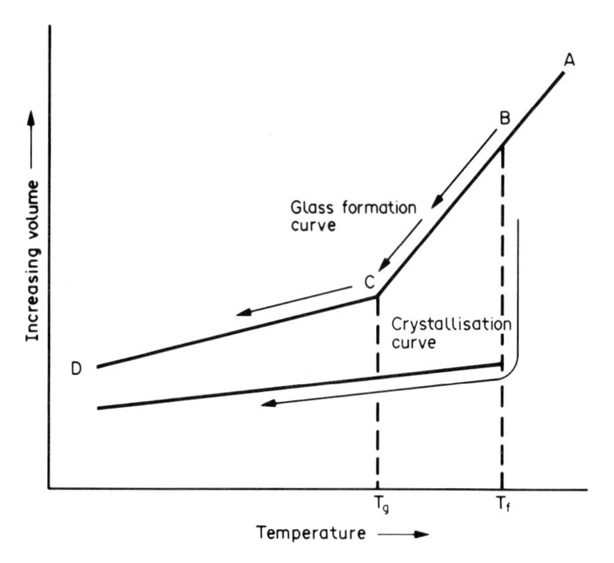

Fig. 6.1 Volume–temperature relationship for liquid, glass and crystalline materials

As the material cools from point A it contracts in volume. If the material crystallises, this will occur at a definite crystallising temperature, T_f (by convention), but if the material "supercools", then there is no definite crystallising temperature. Contraction in volume will, therefore, continue along the line B–C, the shrinkage being partly due to the decreasing range of atomic oscillations and partly due to structural changes in the molecular groupings. These molecular groups in the liquid are restrained by the increasing viscosity; consequently an equilibrium state of crystal formation is difficult to reach.

The viscosity continues to increase with falling temperature and when sufficiently large (about 10^{13}p)[1] at point C, further structural change is prevented. The curve changes slope in the neighbourhood of this point now defined as the transition or transformation temperature T_g. In practice, T_g is not a well-defined temperature and its value depends on the time scale and method of its determination. From this point onwards with further cooling the mass behaves as a solid, a glass. The altering of the cooling rate will alter the temperature at which the curve sharply changes slope, i.e. alter the value of T_g. For this reason, thermal expansion measurements on glaze samples have been arbitrarily standardised in the U.K. ceramics industry at a heating rate of 6°C min^{-1} and values assigned to particular glaze compositions can be discussed between different laboratories.

6.2 THERMAL EXPANSION, MEASUREMENT BY DILATOMETER[2]

Of the common methods of measuring coefficients of thermal expansion the dilatometric technique is the one chosen by most industrial glaze users. Commercially available dilatometers operating automatically measure a complete expansion curve over the whole temperature range from room temperature to softening point.

The most popular apparatus available[3] was originally devised by the British Ceramic Research Association. Although at first it was manually operated, the apparatus can be fully automated.

The fired glaze specimen is heated at a controlled rate of 6°C min^{-1} in a fused silica tube furnace. A fused silica window acts as a stop such that the specimen lies centrally in the heating zone. It is held against the stop by a fused silica rod which in turn is fixed to an Invar rod. Against this is fixed a sensitive measuring gauge. In the manual instrument this is a simple dial gauge, though the alternative is a strain gauge. In the automatic instrument, size changes are recorded via a transducer rather than the dial gauge. A thermocouple projects through the furnace tube close to the glaze test piece.

The test pieces are prepared by pouring the glaze suspension into a refractory mould lined with bat wash or by pressing a rod from the dry glaze powder. Optimum dimensions for the bar are 15 mm × 15 mm × 75 mm. The unfired specimen is melted on the same time–temperature schedule for which the glaze has been designed, followed by an annealing cycle. After firing, the specimen is freed from the bat wash and ground, with coplanar ends, to size. Its length should be measured to ±0.1 mm.

With the specimen in position, the silica push rod is moved up to it, the gauge set firmly against this and the dial zeroed. Each test beings at room temperature and in the manually-operated apparatus the expansion recorded at intervals of 30°C until the transformation temperature is reached. From T_g onwards the expansion is recorded every 5°C until the dial gauge registers its first reverse reading as the glaze reaches its softening point M_g. To these readings is added the quartz correction for that length of sample, giving the corrected expansion. In the automatic instrument the recorder can be adjusted to give a print-out of a silica-corrected percentage thermal expansion curve.

The expansion at any temperature t recorded on the gauge is the increase in length of the specimen, plus the expansion of the silica push rod in the furnace minus the expansion of the outer silica sheath. The expansion of the specimen is the reading on the dial plus the expansion of an equivalent length of fused silica rod. This correction is small and is obtained following dilatometric measurements on a material of known expansion (the standard is usually an alumina bar).

Results can be recorded either as a graph, as a percentage expansion up to a fixed temperature or a coefficient of linear expansion.

1. Graph. A typical curve, taken from a recorder on an automatic instrument is shown in Fig. 6.2.
2. From the gradient of the straight section can be calculated the coefficient of thermal expansion.

The coefficient of linear expansion, α, is expressed as the fractional increase in length when the temperature of the specimen is raised by 1°C. The relationship is:

$$l_t = l_0 \ (1+\alpha t)$$

where l_0 is the length at 0°C,
l_t is the length at t°C.

The coefficient of cubical (volume) expansion is three times the linear coefficient. Values for the transformation temperature T_g and the dilatometric softening point are recorded on the curve.

Percentage linear expansion. The percentage linear expansion between room temperature t_1 and a higher temperature t_2 is obtained from the equation:

$$\text{Percentage expansion} = \frac{(\text{Expansion at } t_2 - \text{Expansion at } t_1) \times 100}{\text{Length } l_{t_1}}$$

By custom in the U.K. ceramics industry the percentage linear expansion is

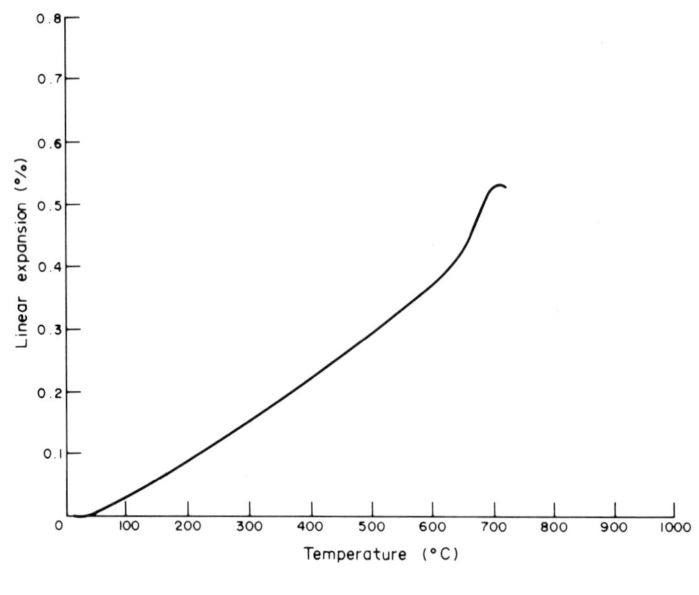

Fig. 6.2

calculated between room temperature t_1 at 20°C and the higher temperature t_2 at 500°C.[4] As a means of routine control, when the expansion of a batch is being compared with that of the standard, comparison of the percentage, linear expansion from 20–500°C is usually adequate. However, as will be explained shortly, the use of such expansion figures can lead to problems in assessing glaze fit between body and glaze.

6.3 GLAZE FIT

The substrate–glaze system is constrained by the ability of the glaze technologist to prepare meltable compositions falling within a narrow band of available thermal expansions. There is a lower limit to the coefficient of thermal expansion above which all practical glazes must exist. The thermal expansion of a glaze is an additive property depending upon the composition,[5] each component contributing to the total expansion by a factor[6a,6b] characteristic of the element. Such factors can be used to calculate a value for the coefficient of thermal expansion, although the figure so obtained is not accurate. In particular, the use of these factors for crystal-containing glazes can lead to marked discrepancies between the actual and the calculated figures. The use of such factors is useful when there is a need to modify glazes which have failed in service.

Stress in the glaze layer develops as a result of a differential thermal expansion between the glaze and the substrate as they cool from the "setting" temperature down to room temperature. The magnitude and sign of this stress determines whether the glaze on the ware will craze, will be serviceable or will peel.

If we consider, in Fig. 6.3, a ceramic body coated with glaze at the glost temperature with the glaze-forming reactions complete, the two components have the same dimension, and any size change in the ceramic has been accommodated by viscous flow in the glaze. As the ware moves through the cooling cycle the glaze begins to "solidify" until a temperature is reached where the two parts, solid glaze and body, are rigidly bonded.

At this point the effect of any difference in the linear coefficients of thermal expansion of the glaze, α_g and of the substrate, α_s, will become apparent. The dilatometer test measures the coefficient on heating, although the stress in glaze–body systems develops on cooling. However, the thermal expansion of the fired body or glaze to a limiting temperature matches the contraction in magnitude and is often called reversible thermal expansion. (Heating an unfired body in a dilatometer results in an irreversible thermal expansion measurement).

We can consider firstly the effect of different coefficients, ignoring for the moment a consideration of the shape of the expansion curves and any possible after-expansion of the body.

1. *Assuming equality of coefficient* $\alpha_g=\alpha_s$. Both glaze *G* and substrate *S* will contract in unison and no strain will be generated.

2. *Assuming* $\alpha_s>\alpha_g$. If the two layers were not constrained by their interfacial adhesion they would contract at different rates. The action of sealing the two together obliges them to be equal in length and stress develops in *both* components. With further cooling the substrate will try to contract the greater amount, but at the end of the cooling cycle the final lengths must be equal. As a consequence the glaze will be compressed and the substrate will be permanently stressed in tension. The tensile stress exerted by the glaze might be great enough to bend the substrate if only one side is glazed. For wall tiles to be flat after glazing it might be necessary for them to be given an initial reverse curvature.[7]

When the difference between the thermal contraction of a substrate is much higher than that of the glaze, excessive compression is generated in the glaze. This can be greater than the strength of adhesion between glaze and substrate and there is then a stress-relieving fracture at the interface known as "peeling". The strength of the body or the interfacial layer has an important bearing on the development of "peeling". A thin or weak substrate can shatter if it is subjected to stress beyond its strength.

3. *Assuming* $\alpha_g>\alpha_s$. As cooling continues, the glaze, which has the higher coefficient of thermal "contraction" will try to contract the more. Again the final lengths of the two parts must be equal so the glaze will compress the substrate and thus must carry a balancing tensile stress. If its tensile strength is adequate, and the glaze is free from flaws, it might be able to withstand the stress. Unfortunately glass is less able to withstand tensile forces than compressive force and even when α_g *is only slightly greater than* α_s the glaze fails in tension with

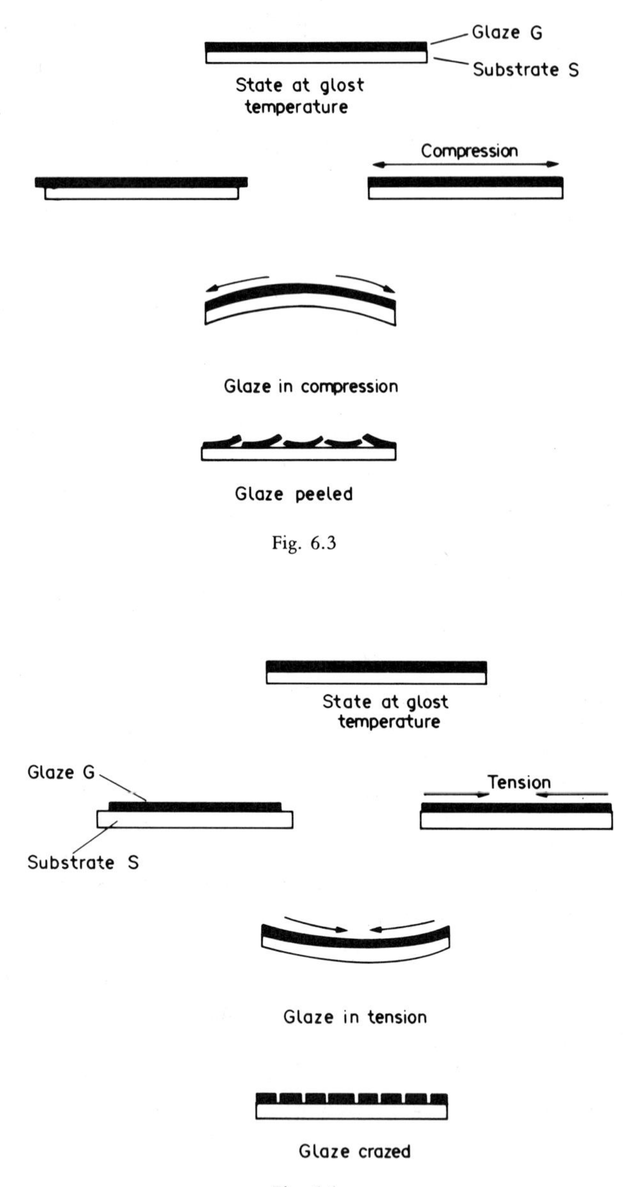

Fig. 6.3

Fig. 6.4

"crazing" being the result. The elasticity of the glaze and the presence of flaws, as well as the coefficient of thermal expansion have an important bearing on the occurrence of crazing.

The induced compressive stress might be great enough to bend the substrate if glazed only on one side.[8] A weak body might shatter rather than respond to this degree of stress.

6.4 THERMAL EXPANSION PROFILE

It is rare that we have a fully-formed vitreous glaze on a ceramic substrate during the rising part of the glost heating cycle. As an aid to understanding events we therefore consider the glaze–body relationship while cooling; that is, as the glaze contracts. When assigning values to the differential thermal expansion we must consider the thermal expansion properties of the glaze as a fully-formed material.

We have seen earlier that as the temperature of a fired glaze rises there are usually associated changes in dimension which, in the majority of vitreous compositions taken to a limiting temperature, are reversible. This is seen in the characteristic thermal expansion curve of a glaze taken to its softening point (but not beyond).

For most glazes in general ceramic use, the dimensional change from near room temperature up to a point near 500°C is positive and linear and remains so to the transformation range. From this zone the glass expands at a greater rate until a point is reached where the specimen apparently contracts.

The several fixed points on the thermal expansion curve of a glaze are determined by its composition and by the heat treatment it receives.[9] Glazes which have crystallised, or devitrified, can have different fixed points from their glassy parent. Some glass-ceramics have zero thermal expansion.

Figure 6.5 shows typical percentage linear expansion curves on the same leadless glaze specimen bars which have been (a) cooled slowly and (b) cooled rapidly from the glost temperature. Some 20–50°C above temperature T_g the curve develops a negative slope due solely to the experimental technique. When the viscosity falls to about $10^{7\cdot 7}$ poise the glaze bar softens to a point where it cannot act as a rigid bar; consequently, it deforms under the small force used to keep the measuring "gauge" in contact with the specimen. This temperature is identified as the (dilatometric) softening point, M_g.[10]

Again it is emphasised that the specimen bar upon which the thermal expansion is determined must be prepared in the same heating cycle which the glaze follows in the glost fire.

On the curve (Fig. 6.5) of the thermal expansion of the typical leadless glaze, the fixed points which are relevant to the question of glaze fit are:

(a) the dilatometric softening point, M_g,
(b) the set point, T_x,
(c) the transformation temperature, T_g, (or transition temperature).

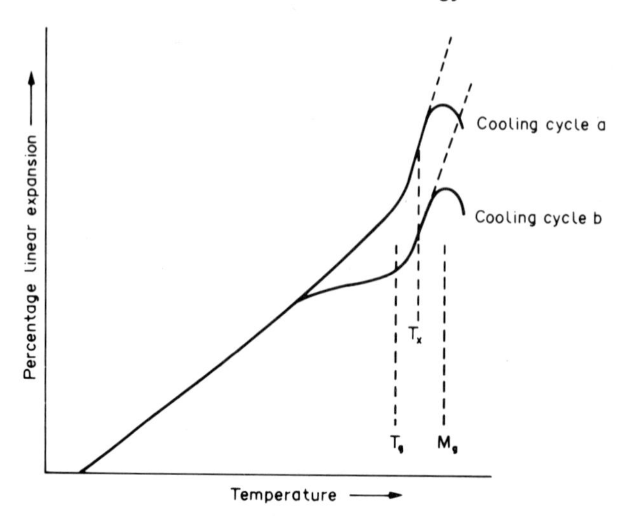

Fig. 6.5. Thermal expansion (a) normally cooled glaze bar (b) rapidly cooled glaze bar.

6.5 DIFFERENTIAL THERMAL EXPANSION, Δ, AND CRAZING RESISTANCE

To help in relating the expansion characteristics of glaze and the crazing resistance it is customary to compare the percentage thermal expansion values of glaze and substrate at 500°C. According to the magnitude and sign of the differential, the liability to crazing or peeling can be forecast to some degree.

While the glaze is above its softening point stress will be relieved, but as the glaze–body composite cools down a new condition develops from temperature T_x onwards. At T_x the glaze behaves as a solid and any stress induced in the glaze from this temperature down will remain. Continued cooling from T_x creates strain in the glaze–body system of a value proportional to the differential contraction between glaze and body to room temperature.

Unfortunately, there is no agreed single position of T_x on the thermal expansion curve. The temperature at which a glaze becomes rigid enough to accept strain has been fixed variously at

(i) $\dfrac{M_g+T_g}{2}$

(ii) T_f (see Section 6.1)

(iii) T_g

The transformation temperature is difficult to define precisely from the expansion graph and a compromise is made by recording the intercept of the produced lines of the two principal parts of the expansion graph. Provided that

TABLE 6.1. PRACTICAL GLAZE–BODY SYSTEMS

Body	Biscuit fire (°C)	Glaze	Glaze firing temperature (°C)	Typical glaze thermal expansion coefficients $\times 10^{-7}$°C^{-1}
"Continental" porcelain	1000	Clear. Lead free. No frit.	1350–1400	50
Bone china	1235	Clear. Medium lead or sometimes lead free. High frit content.	1020–1100	85–100
Vitreous hotel ware	1250	Clear. Low lead. High frit content.	1050	60–70
Vitreous sanitary ware	Single Fire	Opaque. Lead free. Little or not frit. Often coloured.	1250	60–70
Stoneware	{1000 / Single-fire	Various. Lead free. No frit.	1230	65
Semi-vitreous ware	1250	Clear. Low lead or lead free. High frit content.	1050	70
Wall tile	1100	Various. High lead to lead free. High frit content. Often textured.	980–1060	70–80
Floor tile	1180	Some glazed, therefore nonslip abrasion-resistant glaze or engobe needed.	1000–1100	60–70
Earthenware	1100–1150	All types containing frit.	980–1080	70–80
Brick	{950–1120 / Single-fire	All types containing frit.	950–1120	55–75

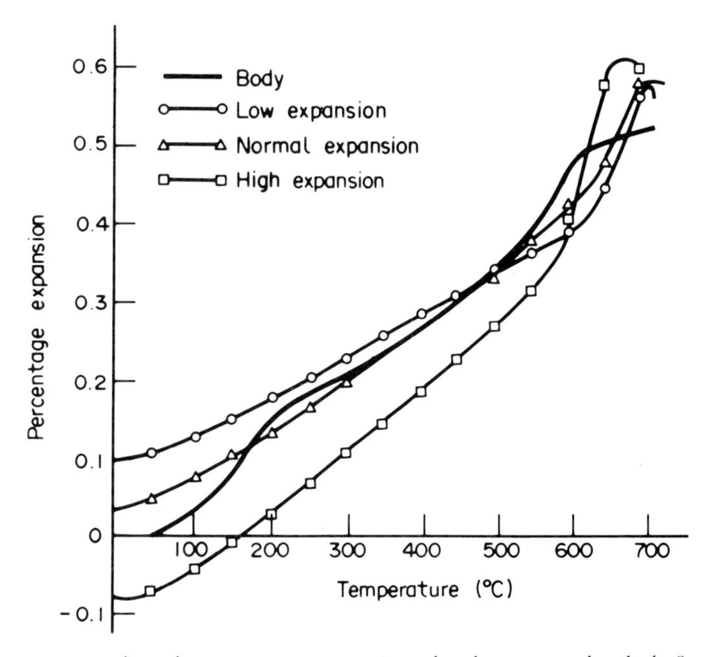

Fig. 6.6. Thermal expansion curves repositioned to demonstrate glaze-body fit

the same convention is always used by the glaze technologist in the same circumstances, then a meaningful picture will exist for stress development in the glaze–body composite. It is proposed that we consider

$$T_x = \frac{M_g + T_g}{2}.$$

The curves of the expansion of glaze and body are drawn to the same scale and considered in conjunction. The graph is moved for the glaze along the ordinate axis over the body graph until the set point T_x of the glaze coincides with a point at the same temperature on the body curve. Depending upon the thermal expansion of the glaze, that is the gradient of the graph, the origin of the glaze curve is displaced by an amount Δ, the contraction mismatch. Any alternative glaze with different thermal expansion characteristics or with a different value of T_x will have a different contraction mismatch.

The behaviour of three glazes on the same body can be examined as examples of this. The set point of each glaze, T_x (where it is assumed the glaze accepts stress), is said to be the half-way point between the transformation temperature T_g and the softening point M_g. In any one family of glazes this temperature will not vary greatly and in the chosen examples set points are shown in Table 6.2

TABLE 6.2

	Glaze	Set point
1.	B 4099	670°C
2.	B 3905	670°C
3	B 3662	620°C

It is this set point which is marked on the glaze curve.

6.5.1 Normal Expansion Glaze

Combined in Fig. 6.6 are the curves for the percentage thermal expansion of an earthenware body and of a leadless glaze designed for this body and known to give good service. Using the ordinate lines as a guide, the marked set point on the glaze curve is superimposed on the body curve. It is noticed at once that the origins of the curves, being displaced and on the same scale, differ by an amount 0.X%, characteristic of this combination *at room temperature.* The difference Δ between the two origins as they cut the ordinate are given in Table 6.3 where the figure is compared with the differences in percentage expansion at 500°C.

TABLE 6.3. GLAZE—BODY DIFFERENTIAL EXPANSIONS

	% linear exp. at 500°C	By difference at 500°C Δ	By set point Δ at room temperature	Type of glaze
Earthenware body	0.34			
B 4099	0.30	+0.04	+0.03	Normal expansion
B 3905	0.235	+0.105	+0.10	Low expansion
B 3662	0.355	−0.015	−0.075	High expansion

6.5.2 Low Expansion Glaze

Normally a glaze with these properties would be used on an oven-to-table body having a lower expansion than the chosen earthenware body. Again the set point has been calculated and marked on the glaze expansion graph. Using the ordinate line as a guide, the glaze curve is moved over the body curve until the set point of the glaze coincides with the same temperature on the body graph. Like the case above, the origin of the glaze curve is displaced from the origin of the body curve and by a greater amount than shown by the glaze with an average thermal expansion.

6.5.3 High Expansion Glaze

Using this technique with the curve of a high expansion glaze, the absence of any compressive stress at room temperature can be forecast.

Practice has determined general guide lines[11] needed for protection against eventual crazing. For more practical purposes an adequate margin of safety is given if the percentage linear expansion of the body exceeds that of the glaze by an amount between 0.02% and 0.06%.

For vitreous bodies the differential Δ lies in the range 0.02–0.04%.

For porous bodies the differential Δ lies in the range 0.04–0.06%.

6.5.4 Thermal Expansion of Crystalline Glazes

Assuming that the glaze at the peak firing temperature is wholly glassy, then as the glaze passes through a deliberately lengthened cooling cycle, crystals form progressively.

As crystallisation proceeds, phases develop (for example of zircon or zinc silicate), which have different thermal expansions from that of the parent glassy glaze. The expansion of these phases can be either higher or lower than the parent depending upon the composition of the crystal phases. Crystallisation will alter the composition of the residual glass phase to a degree depending on the completeness of the new phase development. Consequently the expansion of the residual glass will progressively alter. The effect these changes can have on the thermal expansion of the crystalline glaze can be either:

(a) a counterbalancing of a high expansion phase and a low expansion phase having little resultant effect on the coefficient, or

(b) an imbalance of the phases leading to a coefficient greatly different from that of the parent glaze.

Unfortuntely the eventual effect cannot be accurately predicted.

Again it is emphasised that glaze specimens for thermal expansion measurement must be prepared according to the production glost cycle.

6.5.5 Glaze–body Interface

In considering the development of stress between glaze and body which have different thermal expansions, it has been convenient to think of it as a two-part body–glaze system. Although this is a useful simplification, in practice any problem of crazing or peeling needs to consider the two modes in which the composition of the glaze (and, therefore, its thermal expansion) can change during firing.

1. Fluxes can be lost from the surface by volatilisation.
2. The glaze might dissolve some of the substrate and form a layer of intermediate composition.[12]

In many instances there is sufficient time at the elevated temperature of the glost fire for an interaction between glaze and substrate to produce an interlayer or buffer zone. When molten, the glaze attacks and dissolves the substrate, in some cases leaching selectively and causing a roughened substrate surface. Diffusion of the dissolved elements into the glaze occurs slowly, forming a compositional gradient.[13] The dissolved body material may diffuse through the total thickness of the glaze to the outer surface. As the glazed unit cools, crystallisation in the interfacial zone is initiated, and "layers", whose coefficient of thermal expansion is different from that of either glaze or body,[14] develop. The evidence for such a compositional gradient can be found in an examination of changes in refractive index across the glaze from surface to body.[14]

Factors controlling the development of this zone will be:

1. the composition of the body,
2. the composition of the glaze,
3. the glaze thickness,

4. the maximum temperature reached and whether there is any soaking period during the firing cycle,
5. whether the production technique includes once- or twice-firing.

The in-service behaviour of the ware is affected by the extent and properties of this buffer zone. Trials to test the crazing of glazes fired at different temperatures, for different lengths of time or on non-standard bodies from those normally used, are misleading.

6.5.6 Glaze Stress

The ability of glazed ware to resist delayed crazing is related to the degree of glaze compression. When in compression, glazes increase the mechanical strength and the thermal shock-resistance of the ware. When the stress in the glaze approaches zero, the body strength is reduced. Excessive compression can lead to glaze peeling, but there are other reasons which can account for this effect and knowledge of the stress in the glaze can be useful in identifying and correcting the fault. Several methods have therefore been developed for assessing the magnitude and sign of stress in glaze.[15,16,19,20]

Some are subjective, as is the simple impact test.[17] When the ware is sharply struck by a metal point, the development of concentric cracks around the impact suggests that the glaze is in compression. If the cracks are star-shaped the glaze is probably in tension.

Ring test.[18] This method gives either a qualitative or quantitative assessment of glaze stress. For the test, hollow cylindrical rings 50 mm (2 in) diameter are glazed on the outside only. They are fired at the glost temperature. When cooled, datum lines are marked in one edge of the ring from which measurements are taken.

The ring is cut between the marks with a diamond saw and any stress in the system becomes obvious by movement.

If the glaze is in tension the gap opens, but if it is in compression the gap closes. From these changes in dimension a value for the maximum stress can be calculated. Unglazed blank rings are given a similar treatment and act as standards for the behaviour of the substrate.

The test can be used to measure the effect of moisture expansion.

Tuning fork test. The basis of the test has been long established.[21] Once control is exercised over the preparation of the test specimens the method produces a good laboratory tool for studying the development of glaze stress. A test piece is fabricated from two pieces of body, fixed at their base with slip to a third piece of the same body. After biscuit firing the "fork" is sprayed with a uniform coating of glaze on the outer faces. It is fired according to the glost schedule and annealed. Movement in the tines of the fork is a measure of the magnitude of the stress inherent in the glaze and this movement can be measured by a telescope fitted with a micrometer eyepiece. A stress development curve can be obtained for the glaze by shutting off the furnace at the maximum temperature and recording the readings of the deflection of the tines against temperature.

Microscope method. Optical techniques can determine the type of strain which exists in glaze on its base. Transparent glazes, free from undissolved mill additions or pigment grains, are particularly suited to such methods. It is an advantage if the glaze is applied thickly.

When glass is strained it becomes doubly refracting and if viewed in polarised light between crossed Nicol prisms it shows interference fringes and colours. Strain-free glass is isotropic.

A section about 0.5 mm thick is cut through the glaze and body at right angles to the glaze surface. This is thicker than a ("thin") section cut for mineralogical examination.[22] The surfaces are finely ground with great care to avoid chipping the edges. The glaze face can be protected by resin.

It is mounted on a microscope slide and a cover slip cemented over it using extra resin to protect the glaze. The section is viewed with a sensitive tint plate mounted in the microscope such that the glaze surface is parallel with the "slow" direction. With glaze in compression the section appears yellow. With glaze in tension the section appears blue.

The strength of this colour depends upon the path length of the light through the glaze (i.e. its thickness) and the amount of strain in it. With a known glaze thickness and using a quartz wedge, the amount of the strain can be measured.

The difficulties in precisely defining the "setting" point of a glaze, and with it deducing from two separate expansion curves the fit of glaze and body, can be avoided by directly measuring glaze–body fit with a tensometer.[23,24] This instrument has a horizontal tube furnace closed at one end. At this end there is a support for the test specimen whilst at the opposite end of the instrument, and in the line of axis, there is a system for measuring small movements. A pyrometer is accurately positioned over the centre of the sample to register the standard rate of temperature rises. The apparatus can be used up to 1300°C.

Specimen bars are cast with a trapezoidal cross-section to a uniform weight and size[25,26] from the subject body. This might be biscuit fired before receiving a precisely positioned uniform (but partial) coating of glaze. The specimen is clamped in the furnace. Differences in thermal expansion between glaze and body cause the bar to deflect from the zero datum. The deflection is recorded. A series of mirror deflections is recorded as stress is induced during the cooling cycle.

The setting point of the glaze is given by the point at which the deflection changes from compression to tension.

The test can be carried out as an imitation of the once fired process to determine the effect of glaze on clay bodies. There is a risk of error from possible residual stresses in the body. Autoclave tests on the same combination of glaze and body can confirm the interpretation of the tensometer test results. For routine control of glaze, a stock of standard body is retained.

The effect of changes in firing schedules on the glaze–body interactions can be measured by this method.

The curvature caused by dissimilar thermal expansivity can be measured by transducer.[27] In this method using the cross-over temperature (where the glaze goes into compression on the cooling cycle) rather than the strain-point temperature, leads to correlation between theoretical and practical values of stress-induced camber.

A simple instrument[28] measures glaze fit by recording the differential curvature between glazed and biscuit bars. The instrument consists of a three-sided frame with pins for registering precisely the test ceramic bar. A dial gauge passes through the frame to measure the changes in curvature. The differences in curvature of the bar before and after glost firing is a measure of the glaze fit. Convexity on the glazed side implies compressive stress in the glaze. Concavity on the glaze side implies tension in the glaze. A modification[29] to this technique measures the change in curvature of a glazed specimen when the glaze is ground away.

6.5.7 Crazing Test

Three methods of assessing crazing potential are used;

1. immersion in a steam autoclave,
2. heating in air and quenching in water,
3. heating in steam and quenching in water.

Moisture expansion takes place in porous bodies, rapidly at first on newly fired glost ware but then at a decreasing rate. If this expansion equals the value of the differential thermal expansion Δ, the compression in the glaze will be nullified. When it exceeds Δ crazing will occur as the induced stress overcomes the tensile strength of the glaze. The main practical test for crazing resistance imitates in the short term this long-term behaviour.[30a,30b] Steam under pressure accelerates the moisture expansion of porous ceramic bodies and, in a few hours, test moisture expansions equivalent to years of service can be reproduced.

In the standard test, six specimens of the glazed ware are repeatedly subjected to an atmosphere of steam, in an autoclave. In the autoclave the ware can be subjected to steam at high pressure (340 kN m^{-2}) for hourly periods. To make the body accessible to the steam three small scars (about 6 cm^2) are ground in the glaze to penetrate the glaze–body interface. After each cycle the ware is examined for signs of crazing by rubbing the glaze with a pad soaked in methylene blue dye solution. That which is uncrazed is put through a repeat cycle until a maximum of 16 cycles is reached. Each piece should be exposed only to steam and not to water. With this test thermal shock is avoided.

The number and description of the specimens failing after each cycle is recorded. Service conditions are variable and the differences in the properties of any type of glazed ware are many. There is no direct equivalance between any test attainment and a period of freedom from crazing, but regular crazing tests are valuable.

In Section 6.5 the differential expansion, Δ, was deduced for conditions at room temperature, but there is a valid argument[31] for deducing Δ at the temperature reached by the ware in the "crazing pot", i.e. at 40 psi. This is at 142°C.

Heating in air avoids the effect of moisture expansion and the test results are influenced rather by the shape, thickness and overall dimensions of the pieces and upon the magnitude of the temperature drop on quenching. Because this is a critical feature, variations in the time between removing the ware from the oven and quenching and the speed of quenching are sources of error.

A test[32] was developed which relates thermal shock results to actual working conditions and possible life span in service. Initially the test piece is heated to 120°C in an electric oven, then plunged into cold water at 20°C. If no crazing can be detected the cycle is repeated but with an increase in the temperature of 10°C. Each heating is gradual to ensure the article reaches the required temperature. Successive temperature rises and cold water quenches are made until the glaze crazes.

Criteria were established relating the crazing temperatures to the probable life in service.

Quenching limit of 120°C–possible crazing in a few days.

Quenching limit of 180°C–ware uncrazed for 2 years.

This test is not often used because its results are difficult to assess. With moisture expansion and thermal shock combined in one test the factor causing crazing cannot be stated.

6.6 GLAZE–BODY REACTIONS (BUFFER LAYER)

There is a wide choice of "recipes" for any given glaze composition. For any fritted glaze there can be a combination of low-temperature borax frit, lead bisilicate, felspar and china clay or it could be a combination of lead borosilicate frit, felspar and clay. The rapidity with which the frits lose viscosity and gain in reactivity will be different in each case.

Once a liquid phase forms, attack on the substrate begins, leading to the formation of intermediate compositions which could be either vitreous or crystalline. The mechanism of this corrosion is similar to acid–base reactions in aqueous solution.[33] There is subsequent diffusion of chemical species from the body into the glaze and from the glaze into the body. A well-developed reaction zone develops and influences the ability of the glaze to resist imposed stresses. Once the zone is formed the reactions slow down.

These intermediate compositions can be expected to have different viscosities and surface tensions and consequently have different powers to release bubbles. The same glaze fired on a variety of bodies can give quite different surface qualities because of this factor. The difference can be attributed to porosity in the substrate, in some instances.

During slow-cooling, crystals form which can extend from the unreacted body into the glaze. Fast cooling prevents crystal growth and the interface can only be seen as an amorphous layer. Mechanical strength of the ware can be affected by the magnitude of this crystal layer and by whether it retains an adequate bond with the body. The reaction zone can comprise 25% of the total original glaze thickness.

Evidence of this crystal layer can be seen in the cross-section of a glaze–body interface. A fully-fritted, low-solubility glaze, compounded from a lead borosilicate frit (ground with 1% clay) fired onto bone china shows three distinct layers when scanned by electron microscopy. There is an intermediate composition between glaze and body which can be seen in the illustration of a transverse polished section of this system. The boundaries are sharp. Element distribution images can be prepared for the same area. These images (seen in Fig. 6.7) show the relative concentration and distribution of silicon, aluminium, phosphorus and lead through the body–buffer–glaze interface.

In the images, the density of the white dots indicates the abundance of a particular element. From the illustrations it is evident that there is more phosphorus in the body (as the component calcium phosphate) than in the glaze and that the crystalline interface resembles more the body itself for phosphorus content. Diffusion of lead from the glaze into the body is very evident.

The "raster" picture can be supplemented by a single line scan of elemental concentration from point to point. X-ray intensities along the line in the photograph are displayed in Fig. 6.8. In the case of lead the concentration falls away to zero at some distance into the body. At the glaze/air surface there is an abrupt change in the Pb trace and whilst this could be construed as a loss of lead by volatilisation it could be due also to either a loss of signal from a curved surface or to an overlap of the beam falling outside the glaze.

TABLE 6.4 X-RAY INTENSITY MEASUREMENTS
AVERAGE INTENSITY (COUNTS/10 SEC)

	Surface	Centre	Body–glaze interface
AlK	1179	1240	1400
SiK	7500	7366	7243

The aluminium trace shows a significant, steady decrease through the glaze. Detailed measurements (in counts per 10 sec) in Table 6.4 show a decrease of nearly 20% from the interface to the surface. A similar examination for silica shows the reverse trend with a difference of about 5% from surface to interface.

The nature of the body composition is significant in the way it influences subsidiary glaze reactions, such as the manner in which it affects the bubble population.[34] High-quality gloss cannot be developed in a glaze either still containing, or which has recently contained, much bubble. Bodies containing materials processed by hydrocylone generally have been found to give a glaze with an improved surface.[35] The quality of glaze is related to the size and amount of free silica in the body and by inference to how many large free silica grains or crystals occur at the glaze–body interface. Bodies with the lowest amount of free silica give the best glaze appearance. Consequently a glaze fired on bone china gives a better finish than when the same composition is used on earthenware.

During maturation of *any* glaze the amount of bubbles present is not so very different, but what distinguishes glazes with poor surfaces from those with good surfaces is the amount of bubbles they retain after firing. Those bubbles which remain are often associated with edges and points of free silica crystalline material.

Dissolved surface components in the body will include iron impurities and other colouring ions (if present) and as these enter the glaze structure the coloured zone is enlarged, magnifying the impact of the original specks or contamination. Fluid lead glazes can become uniformly coloured cream through the overall solution of a high iron (terra-cotta) body.

Green spot is a fault seen in the glazing of fireclay products. The spots, frequently intensely coloured, are caused by the presence in the clay surface of a copper containing impurity ($CuFeS_2$) which subsequently dissolves in the glaze during the glost fire.

Zircon opaque glazes in reacting with the body are modified by the dissolved elements, often to such an extent that the intermediate layer between glaze and body is transparent.

Analysis by an energy dispersive X-ray technique is used to examine the behaviour of the special glazes which coat electro-ceramic substrates. The insulating properties of the glaze must not be modified by unwanted ion migration and compositions can be designed to reduce such ionic to a minimum.

Migration of aluminium and lead ions on firing a single ground lead borosilicate composition (70% PbO, 20% B_2O_3. 10% SiO_2) on a 96% alumina substrate has been demonstrated.[36] Examination of the glaze and body to 6 μm on either side of the glaze–body interface on a sample fired at 800°C for 10 min was carried out by energy dispersive analysis/scanning electron microscopy. This revealed that the PbO content (wt%) within the glaze at 2 μm from the interface was 70% but fell to approximately 40% at the apparent interface. On the body side of this apparent interface, PbO was detected falling from 40% at the

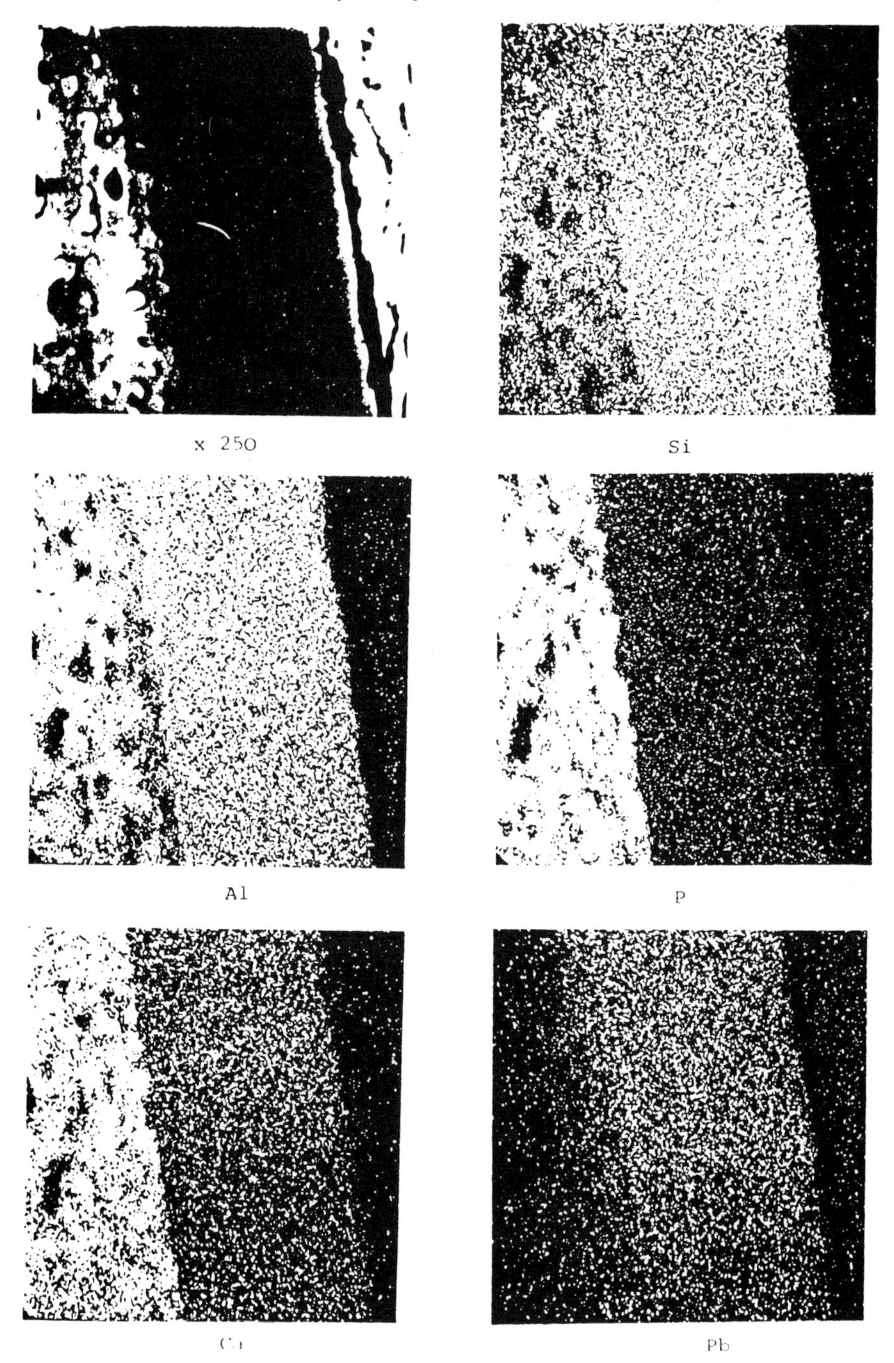

Fig. 6.7: Element distributions

x 350

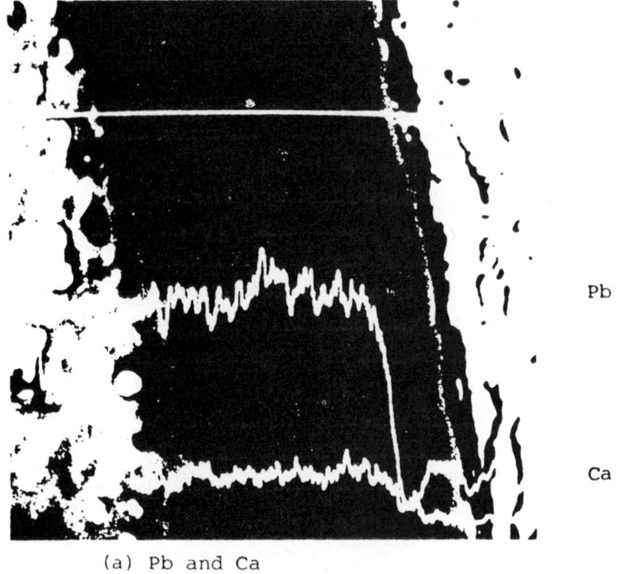

(a) Pb and Ca

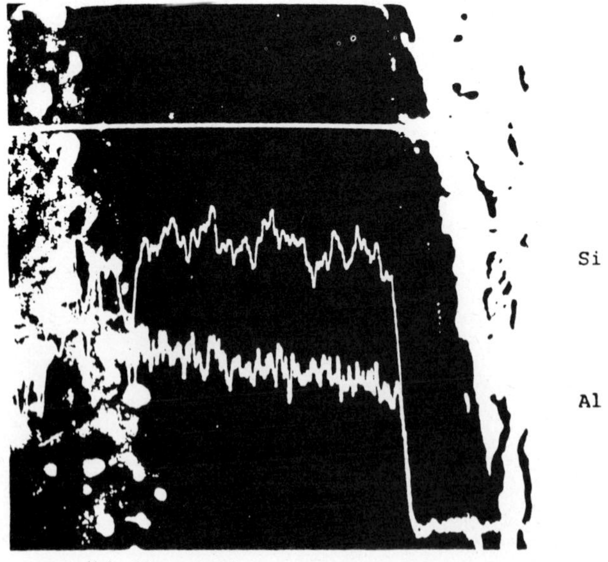

(b) Si and Al

Fig. 6.8: Line scans

interface to zero at 2 μm within the body. The Al_2O_3 content was found to fall from the 96% present in the body at 2 μm from the interface to 60% at the interface and then to level out at approximately 20% at a distance of 2 μm within the glaze from the interface. This Al_2O_3 level was retained from 2 to 6 μm from the interface within the glaze.

The fired glaze thickness was reported as being 11 μm. Such findings demonstrate that the interaction of a thick film glass with the substrate can affect the properties of the "as fired" composition compared with that expected from the theoretical glass composition.

A knowledge of the manner in which ions migrate can be used to explain some aspects of the *surface* behaviour of a glaze. Under favourable circumstances some body ions can move across the thickness of the glaze. Following a change in the flux component of a ceramic substrate, glazed normally with a potassium-free composition, an applied lithographic decoration did not sink into the glaze surface to the degree expected (this would have an adverse effect on the durability of the decoration.) Subsequent analysis showed that K^+ ions from the body had replaced Na^+ ions in the surface layers of the glaze to a degree which modified the glaze viscosity at the decorating temperature thus preventing flow.

REFERENCES

1. *Treatise on Materials Science and Technology*, Vol. 22 Glass, III p. 261 (1982).
2. *Thermal Expansion Test.* British Ceramic Society (1976).
3. Available from Malkin Ltd, Stoke on Trent ST4 4ES.
4. James, W. and Norris, A. W. *Trans. Brit. Ceram. Soc.* **55,** 601–630 (1956).
5. Blau, H. H. *J. Soc. Glass Tech.* **35,** 304–317 (1951).
6a. English, S. and Turner, W. E. S. *J. Amer. Ceram. Soc.* **10,** 551–560 (1927).
6b. English, S. and Turner, W. E. S. *J. Amer. Ceram. Soc.* **12,** 760 (1929).
7. A. T. Green Book, pp. 155 et seq. British Ceramic Research Association.
8. Peychès, I. *J. Soc. Glass. Tech.* **36,** 164–180 (1952).
9. Ed. Tomozawa and Doremus. *Treatise on Material Science and Technology* (Vol. 22) pp. 260–262. Academic Press (1982).
10. Bullin, L. and Green, K. "Crazing of wall tiles". *Trans. Brit. Ceram. Soc.* **53,** 39–63 (1954).
11. Bruce, R. H. and Wilkinson, W. T. *Trans. Brit. Ceram. Soc.* **65,** pp. 233–276 (1966).
12. Thomas, E., Tuttle, M. A. and Miller, E. *J. Amer. Ceram. Soc.* **28,** 52–62 (1945).
13. Perricone, A. C. and Stone, R. L. *J. Amer. Ceram. Soc.* **37,** 33–38 (1954).
14. Smith, A. N. "Investigations of glaze–body layers". *Trans. Brit. Ceram. Soc.* **53,** 219–228 (1954).
15. Bell, W. C. *J. Amer. Ceram. Soc.* **23,** 163–166 (1940).
16. Johnson, R. *Ceramics,* **21,** 16–21 (1970).
17. Wood, F. T., and Hind, S. R. *Trans. Brit. Ceram. Soc.* **38,** 435–454 (1939).
18. Schurecht, H. G. *J. Amer. Ceram. Soc.* **11,** 271–277 (1928).
19. Lacan, B. and Maupoil, J. J. *Ind. Ceram.* No. 660. 181–188 (1973).
20. Steger, W. *Bericht. Deut. Keram. Ges.* **9,** 203–215 (1928).
21. Blakely, A. M. *J. Amer. Ceram. Soc.* **21,** 243–251 (1938).
22. Clarke-Monks, C. and Parker, J. M. *Stones and Cord in Glass.* Pub. Soc. Glass Technol. (1980).
23. Steger, W. *Ber. Deut. Keram. Ges.* **9,** 203 (1928).
24. Cox, D. L. *Ceram. Ind. Journal,* **92,** 24–26 (1983).
25. Lacan, B. and Maupoil, J. J. *Ind. Ceram.*, pp. 181–188 (1973).

26. "Measurement of stresses in glaze". Report No. 16, Deutche Keramischen Gesellshaft. Baden (1962).
27. Young, W. S. and Wilcox, D. L. *Bull. Amer. Ceram. Soc.* **51,** 672–676. (1972).
28. Radford, C. *Trans. Brit. Ceram. Soc.* **76,** xx–xxv (1977).
29. Harrison, R. *Trans. Brit. Ceram. Soc.* **77,** xii (1978).
30a. *Craze Resistance Test for Tableware.* British Ceramic Society (1977).
30b. Crazing Test, ASTM C424–58T.
31. Bullin, I. and Green, K. *Trans. Brit. Ceram. Soc.* **53,** 39–66 (1953–4).
32. Harkort, H. *Trans. Amer. Ceram. Soc.* **15,** 368–373 (1913).
33. Roberts, W. and Marshall, K. *Trans. Brit. Ceram. Soc.* **69,** 221–224 (1970).
34. Franklin, C. E. L. *Trans. Brit. Ceram. Soc.* **64,** 549–566 (1965).
35. Dinsdale, A. *Trans. Brit. Ceram. Soc.* **62,** 321–338 (1963).
36. Machin, W. S. and West R. W. In *Processing of Crystalline Ceramics*, Eds. Palmour, H., Davis, R. F. and Hare, T. M. p. 243. Plenum Press (1978).

Chapter 7
Glaze Formulation

7.1 CLASSIFICATION OF GLAZE

The classifications into which glazes can be put reflect the many characteristics of the glaze itself, the substrate, the final glazed product or some feature of the glaze application or firing process. The following features are attributed either singly or in combination to a glaze by way of classification.

1. Lead or leadless.
2. Raw or fritted.
3. Once-fired or twice-fired. (Once- or twice-fired refers to the substrate: once-fired glazes are applied to clayware and twice-fired glazes are applied to biscuit.)
4. Tile, tableware, sanitaryware, electrical porcelain, etc.
5. Hollow-ware or flatware.
6. Earthenware, hard procelain, bone china, vitreous china, alumina, etc.
7. Dipping, spraying, waterfall, etc.
8. Fast-fire or "conventional" fire.
9. High temperature or low temperature.
10. High expansion or low expansion.
11. Oxidising, neutral or reducing fire compositions.
12. Coloured or colourless.
13. Transparent or opaque.
14. Glossy, matt, vellum or textured, etc.
15. Electrically conducting, scratch-resistant, etc.

Categories 4, 6, 7, 14 and 15 include only a selection of possible examples and other categories reflecting characteristics which might be important in other circumstances could be added.

The above list highlights the complex nature and the interrelationship of glazes, of the substrates to which they are applied and the manner in which they are employed. Of the categories listed, category 1 refers to the chemical composition of the glaze and 2 to its batch composition, 4 refers to the use to which the final product will be put and 5 gives an indication of the shape of the ware to which the glaze is applied. Category 6 refers to chemical and physical characteristics of the fired substrate, while categories 3, 7–9 and 11 indicate aspects of the processing methods employed in the manufacture of glazed ceramic ware. The appearance of the fired glaze is covered by 12–14, while 10 and 15 refer to specific properties designed into the fired glaze composition, often to enable the finished product to perform some required function.[1]

Although individual glazes for a particular substrate can have characteristics in common, glazes for other substrates could be very different in terms of these descriptions. For example, glazes for felspathic (hard paste) porcelain are raw, leadless, usually transparent, colourless and twice-fired, but earthenware glazes can be either lead or leadless, transparent or opaque, glossy or matt, coloured or uncoloured, raw or fritted and once- or twice-fired.

Most of the characteristic descriptions outlined overlap many categories of substrate and these aspects will be considerd initially. Finally, glaze compositions designed for use with specific substrates and/or employed in a specific manner will be dealt with in more detail.

7.2 LEAD OR LEADLESS

Although other chemical characteristics of glazes can be distinguished, differentiation between lead and leadless compositions has always been the most significant. The major reasons for classifying glazes as either lead or leadless are related to hygiene and the quality of the fired surface.

The absorption of lead into the human body has long been known to be deleterious to health, and in the case of inorganic lead compounds those with a higher solubility in the acid environment of the stomach present a greater hazard. Lead poisoning was a danger faced by operators in the U.K. pottery industry during the nineteenth century as a result of the use of white lead, $2PbCO_3.Pb(OH)_2$, in glazes. This material, which suspends well in glaze slips due to its flakey nature, is soluble in acid and consequently a health hazard. Action in the U.K. towards the end of last century resulted in the use of lead melted into glassy frits which reduced the level of exposure to acid-soluble inorganic lead. Later, low-solubility regulations were introduced for glazes and the use of such compositions has eliminated lead poisoning through exposure to lead glazes in the U.K. industry. Low-solubility glazes are ones which release less than 5% of their own weight of a soluble lead compound, expressed as PbO, when a sample of the glaze, in the ground form, is tested under specified conditions. Briefly, these are that after drying at 110°C the sample is shaken with 1000 times its weight of 0.07 M HCl at 23±2°C for 1 hour, allowed to stand for 1 hour before it is filtered. The filtrate is analysed for the presence of lead (Government test).

Lead glazes produced and used in the U.K. include in their recipe either lead bisilicate and/or lead borosilicate frits as the only source of lead. Lead bisilicate as manufactured for use in glaze does not correspond exactly to the composition $PbO.2SiO_2$, but includes a small amount of Al_2O_3 to enable a lower lead solubility to be attained by the frit without an undesirable increase in viscosity. For practical purposes, using lead bisilicate of typical composition (Seger formula)

$$1.00\ PbO\ 0.10\ Al_2O_3\ 1.89\ SiO_2\ \text{(by weight: 64\% PbO, 3\% } Al_2O_3\ \text{33\% } SiO_2)$$

allows glazes of the lowest solubility to be achieved (i.e. the ratio of soluble PbO to total PbO is at a minimum).

Designers of lead borosilicate frits for use in the U.K. must take account of the lead solubility characteristics so that resultant glazes satisfy the low-solubility requirements. Increasing the amount of alkali oxides and B_2O_3 results in an increase in the lead solubility of the frit; nevertheless, lead borosilicate frits with satisfactory solubility levels can be produced for tableware glazes, i.e. for bone china or earthenware, by introducing modest amounts of alkali oxides and B_2O_3 into a lead-containing glass. Routine testing of all lead frits for lead solubility is an essential control test prior to glaze manufacture.

As would be expected, the lead solubility (Government test) of a lead glaze increases as the surface area per gramme of dry glaze increases. Therefore, during manufacture, control of the particle size of the glaze is as necessary for the observance of solubility regulations as it is for other reasons.

Other countries (e.g. The Netherlands) also have statutory regulations requiring glazes to possess specified lead solubility characteristics. However, in some countries less durable lead frits having greater lead solubility in acids can be employed because no solubility restrictions exist. In such countries low melting and highly soluble lead silicate and lead borate frits can be used. This gives greater freedom in glaze formulation such that unique fired glaze effects, which cannot be achieved with a low-solubility formulation, are attainable.

Although few countries have regulations specifically designed to reduce the exposure of pottery workers to acid-soluble lead glazes, many have introduced measures aimed at protecting consumers from the effects of extraction of toxic heavy metals, particularly lead and cadmium, from ceramic tableware and cookware.

These measures restrict the amount of toxic metals which can be extracted by an acid solution under specified test conditions and they were introduced partly as a result of isolated incidents of lead poisoning which occurred after storage of acidic liquids (e.g. fruit juices) in glazed vessels produced without a sound technical knowledge of glazes. For industrially-made ceramic tableware, lead release from lead containing glazes is usually at a very low level, well below the legislative limits. The major source of lead released from ceramic ware is due to the use of onglaze enamels based on low melting lead glasses. (Further mention of lead release will be made in Sections 7.11, 9.5 and 9.6.)

Another reason for differentiating between lead and leadless glazes is related to the quality of the finished ware. Indeed, in view of the recognised hazards involved in the use of lead compounds in glazes, some major advantages would be needed to offset the health and safety disadvantages. As has been mentioned in

Chapter 3, the inclusion of lead compounds in a glaze produces a composition having a wider maturing range and a better capability of healing over faults (due to its lower viscosity and surface tension) than would be the case with a leadless glaze. Lead glazes have greater brilliance than leadless compositions because of their higher refractive index. These benefits have resulted in the continued use of lead glazes for high quality tableware and ornamentalware.

Due to marked volatilisation of lead at temperatures above 1150°C lead glazes are not usually employed if they are to be exposed to temperatures above this level for any length of time. They are, however, employed on earthenware, bone china, vitreous china and other vitreous tableware bodies where peak temperatures of 1000–1120°C are reached in the glost firing cycle.

In Britain, a "leadless" glaze is defined as a composition having a PbO content of less than 1% by weight [Pottery (Health and Welfare) Special Regulations 1950]. Factories using only such glazes, which usually do not contain any deliberately-added lead compounds, can be designated as "leadless" factories. In such factories, the precautions which have to be taken to ensure safe and healthy working conditions in glazing shops are clearly different from those necessary in factories where lead glazes are employed.

7.2.1 Strontium Glazes

Potential hazards exist which are associated with the manufacture and use of lead glazes. There is the possible presence of lead-containing dusts in the factory atmosphere during the production of frits and glazes; lead compounds can appear in liquid effluents and lead ions could be released from glazed ware. Attempts have been made therefore to develop leadless compositions which have the low-melting characteristics, low viscosity, wide firing ranges and the high gloss of lead glazes.

The presence of major amounts of certain oxides in a leadless glaze composition have some desirable characteristics, but at the same time there is a tendency for the glaze to have undesirable properties. High alkali contents result in lower melting temperatures but only at the expense of a markedly increased thermal expansion (Na_2O, K_2O), increased cost (Li_2O) and reduced durability. Increased gloss results from the presence of BaO as glazes containing barium exhibit a high refractive index. However, many compounds of barium are toxic (e.g. barium carbonate) and release of barium ions from the fired glaze is possible on exposure of the latter to acidic fruit juices.

The replacement of lead by strontium has been found to result in the desirable properties imparted by the presence of lead being retained to a significant degree. Much of the current interest in strontium glazes follows extensive studies in the U.S.S.R.[2] Glazes can be formulated which, by comparison with lead-containing glazes, are non-toxic. However, they have not been widely adopted as they are inferior to the quality of lead glazes. Only in the U.S.S.R does it appear that strontium glazes have gained a significant position in the industrial market.

In addition to having a wide firing range, low firing temperature and the ability to develop a glossy surface, strontium glazes have also been shown to exhibit excellent scratch-resistance.[3]

Leadless strontium glazes formulated for firing on semi-vitreous and vitreous tableware bodies have been described.[4] In order to achieve the required thermal expansion characteristics ($6–7 \times 10^{-6}\ ^{\circ}C^{-1}$), the alkali content had to be restricted to 0.19 mole (in the Seger formula), with K_2O being the major alkali so as to give good gloss. In addition, relatively high SiO_2 levels (0.42–0.51 mole) were necessary to attain the desired maturing characteristics in combination with Al_2O_3 (0.25–0.40), SiO_2 (2.7–4.0), B_2O_3 (0.25–0.70) and other alkaline earth oxides, principally CaO. All the compositions were fritted and mill additions of china clay (typically 10–12%) were made to produce the glaze batch. Firing temperatures were typically 1150–1180°C.

Liquid–liquid phase separation, giving an opalescent effect, can occur in strontium borosilicate glazes if insufficient Al_2O_3 is present. Alternatively, excessive Al_2O_3 can result in crystallisation of alumina-rich phases. Strontium glazes are however not generally considered as a means of producing matt effects because these can be achieved more economically by other means. Indeed, the relatively expensive nature of strontium minerals has been a factor in limiting the use of strontium glazes.

When used in conjunction with underglaze colours, leadless strontium glazes have little if any effect on the colour tone observed, but if used in combination with chrome tin pink stains, the glazes should contain sufficient calcium oxide (approximately 0.35 mole minimum) to stabilise the colour.

7.3 RAW OR FRITTED

As raw glazes are compounded without the use of frits, they are generally restricted industrially to applications were peak firing temperatures of 1150°C or above are employed. They are used in the production of hard porcelain, vitreous china sanitaryware, stoneware, electrical porcelain and various glazed low expansion bodies.

Raw glazes are based on mineral fluxes such as felspars or nepheline syenite, with the addition of clays, quartz, calcium carbonate (ground limestone or whiting), dolomite, zinc oxide and zirconium silicate as common components. Petalite is employed as a flux in low-expansion raw glazes.

Raw glazes do not contain any preformed glassy phase (frit) and therefore sufficient time must be available during the firing cycle to allow for the evolution of gasses from the raw material components (e.g. CO_2 from carbonates, water vapour from clays) and for dissolution processes to proceed to the point where smooth, relatively bubble-free glaze can be obtained. Firing schedules will therefore be generally slower for raw glazes than for fritted glazes.

Raw glazes are often applied to ware which is subsequently produced by a once-fire process where the glaze and body mature together. As the glaze contains no frits, it vitrifies and seals over at a relatively high temperature after the evolution of reaction gases from the body materials has ceased.

Fritted glazes are essential if peak firing temperatures below 1150°C are to be employed. Although frits can be employed in glazes firing above this temperature, loss of volatile constituents (e.g. B_2O_3 and PbO) can be appreciable and be the cause of problems.

Fritted glazes usually contain materials other than frits; often only clays will be present, but complex formulations containing a number of frits (lead and leadless) and a wide range of minerals and inorganic industrial chemicals are known. Glazes containing only frit or additionally only a very minor amount (1%) of a suspending aid are described as fully-fritted. Many glazes for use on tableware and tile bodies are fritted.

With increasing interest in the use of fast-firing techniques for the production of tableware and tiles, the trend is towards glazes with an increased frit content because the maturing of the glaze on the ware can be more speedily achieved through the additional glass content. With a fully-fritted glaze, a glass of uniform composition will firstly have been obtained during the fritting process and a fast-firing glost cycle will be facilitated by its inclusion in the composition.

7.4 ONCE-FIRE OR TWICE-FIRE

These terms refer to the number of times that the substrate is fired to give the glazed product and not to the number of times the glaze is fired. (The term "third fire" refers to a subsequent enamel decorating fire.)

At first glance, producing ware by a once-fire rather than a twice-fire process would appear to be the more economic because less energy will be consumed. However, in determining whether once- or twice-firing is more economic, factors other than simple firing costs have to be considered.[5]

Once-firing is favoured when glazing complex-shaped large articles (e.g. sanitaryware or insulators) of high value. For such articles it is not possible to achieve a high setting density in the kiln and therefore use of a twice-fire process would be expensive both from the kiln fuel consumption viewpoint and from the need to store ware at the intermediate biscuit stage. These large articles are not completely covered with glaze and they can support themselves without distortion during the once-firing process.

For tableware, in particular, and tiles, twice-firing is often favoured because high density settings can be achieved in biscuit kilns, especially in the case of porous ware (earthenware flatware, tiles) and in glost kilns. Additionally, storage of the low-value biscuit items is not too expensive because many of the shapes have high storage density. This intermediate storage stage aids production flexibility particularly where many shapes are manufactured and orders can be processed from this stock rather than making from the clay stage. A major advantage of twice-firing is that selection at the biscuit stage can be carried out and defective products eliminated at a point when the costs incurred are relatively small.

From the technical viewpoint, twice-firing is necessary in certain cases where individual items have to be supported during a biscuit fire (e.g. bone china flatware) to prevent distortion. In addition, glazing biscuit articles can be more readily achieved and controlled than when glazing clayware, leading to improved quality and reduced losses.

The advantages which accrue from twice-firing can outweigh the additional firing costs which are incurred compared with the case of the once-firing. The situation is not static and a move towards the increasing use of once-firing techniques can be anticipated.

In a once-fire process, the glaze will be designed to mature at the same time as the body. However, the vitrifying glaze must not seal completely before gaseous evolution from the body interactions has ceased unless these gasses can escape through an open body pore system to an unglazed surface, as is the case with porous tiles, which are glazed on one face only. Premature sealing-over of the glaze can lead to a sub-standard, pinholed, surface. When body and glaze mature together, an intermediate layer develops which usually gives added strength to the ware. This is particularly so when the body fully vitrifies.

For a twice-fired product, the glost temperature is usually lower than that reached in the biscuit firing and consequently no further contraction of the body occurs. This is the usual situation with twice-fired china and earthenware and it allows the placing and firing of flatware in cranks and a relatively high setting density in the glost kiln. Felspathic porcelain differs in this respect being first subjected to a low-temperature biscuit fire (900–1000°C) to produce a rigid porous body. This, after glaze application, is fired to a very much higher temperature (1400°C) resulting in simultaneous body vitrification and maturing of the glaze. For procelain flatware, the use of individual setters in the glost fire is accepted practice. Earthenware can also be produced by an easy biscuit/hard glost process.

With the increased interest in fast-firing for tiles has come a greater awareness of the benefits which can accrue from using a once-firing process. Being flat and glazed on one side only, tiles are particularly suited to a rapid-once firing production process.

Where once-firing techniques are employed, it is usual to include in the glaze slip a binder which will both control the rate of removal of water from the slip and its movement into the extremely porous body. In addition the binder imparts hardness to the glaze surface as drying takes place, thus facilitating handling the ware prior to firing.

7.5 HOLLOW-WARE OR FLATWARE*

In many areas of tableware production, it is necessary to employ different glazes for hollow-ware and flatware even when the two products are glost fired under identical conditions. Where lead glazes are employed there is a slight flow down vertical surfaces and it is for this reason that hollow-ware glazes are often designed so that they are somewhat more viscous at peak glost temperature than the corresponding glaze used on flatware.

7.6 FAST-FIRE[6] OR "CONVENTIONAL" FIRE

These terms relate to each other, but do not themselves imply any particular length of firing cycle. What is considered a fast-firing cycle today or at any time in the future may, 5, 10 or 20 years later, be thought conventional.

Following the rapid, steep rises in the costs of fuels during the 1970s, attention has been increasingly directed towards making savings in the costs of firing ceramics. As a result, low thermal mass fibre-lined intermittent kilns have become very popular and these allow a more rapid firing schedule to be employed. The new schedules have generally been achieved without the need to reformulate the glaze composition. A typical gain from the use of this type of kiln has been the reduction in the glost firing time for bone china to between 5 and 8 hours from cycle times of 24 hours or more in brick-lined tunnel kilns. In addition to these new intermittent kilns, low thermal mass roller hearth and other types of "tunnel" or passage kilns have been introduced. These kilns permit glost cycles of 45 minutes or less to be achieved for suitable substrates. Once-firing of tiles on cycles of less than 2 hours is feasible.

At the time of writing, fast-firing for tiles and tableware is generally taken to imply a cycle of 3 hours duration or less. For large complex pieces such as sanitaryware, a 10–15 hour once-firing schedule is presently considered to be fast.

Glazes for fast-firing on biscuit earthenware and china generally require a greater proportion of frit to be included in the glaze batch then has hitherto been the case if cycles of 3 hours or less are to be achieved. For vitreous bodies such as bone china, a completely fritted system can be employed although for porous earthenware, use of such a glaze can lead to blistering. However, this fault can be alleviated by the inclusion of an amount of clay in the glaze mill recipe.

Rapid firing glazes for tiles can also be formulated with a high frit content. For crystalline effects, it is generally necessary to include a suitable non-fritted material in the glaze to give the desired surface texture because the rapid cooling of the ware during a fast-firing cycle does not allow time for those crystallisation processes which occur when slower cycles are employed.

* Flatware items include saucers and plates; hollow-ware articles include cups and mugs.

The success of producing ceramic ware by fast-firing techniques is influenced to a major extent by the properties of the substrate. The presence of phases within the fired substrate which undergo crystal inversions involving volume changes during glost firing is detrimental if fast-firing techniques are employed. Dunting in the ware is a likely result of these inversions. High thermal expansion bodies and substrates of large dimensions also suffer from dunting if fast-firing techniques are employed because of the stresses set up due to uneven cooling (or heating) over individual articles.

Standard earthenware bodies contain, after firing, appreciable amounts of cristobalite and/or quartz, both of which undergo crystalline inversions involving major volume changes at temperatures of approximately 200°C and 573°C respectively. Consequently, it is not possible to fire earthenware as rapidly as a substrate with a uniform and a lower thermal expansion, other factors being equal. When considering fast-firing techniques for the production of glazed ceramic ware, reformulation of the substrate composition may be necessary as has been done in the tile and tableware industries. Fast-firing can alter the properties required in a suitable glaze.

7.7 COLOURED GLAZES

Although many different glaze effects can be achieved without the use of materials which impart colour to the final product, employing such materials gives greater freedom to designers. Although some ceramics are purely decorative, most are functional. However, the manner in which these are used is such that their aesthetic appeal is of considerable importance to the potential purchaser. Colour therefore is critical to successful sales.

With coloured glazes it is necessary to differentiate between the industrial user and the craft or hobby potter. The craft potter rarely requires to reproduce indentical products—indeed consumers of craft products are attracted by the individuality of each piece made. However, the mass-produced industrial products must be manufactured such that designs of shape and colour can be reproduced over a number of years to an agreed standard. This is particularly true of sanitaryware articles and tiles. Wall tiles of a particular design have to be reproduced to a standard because they are used in situations where each tile must match those which surround it in an installation.

It is therefore necessary for the industrial producer to ensure that coloured glazes match from batch to batch. This need has led to the widespread use of stable ceramic pigments which are manufactured by specialist producers.[7] Compared with metal oxides and other colouring minerals, these manufactured pigments are more consistent in their colour properties. The inconsistency of colouring metal oxides in glazes is not a major drawback to the craft potter and as colorants they are cheaper than manufactured pigments. Their inferior consistency does not preclude their use by industrial users but when so employed they are often constituents of glaze where a "craft" type effect is imitated.

7.7.1 Colorants

Inorganic compounds of the transition metals vanadium, chromium, manganese, iron, cobalt, nickel and copper are generally coloured and the raw oxides, and Sometimes carbonates, of all these elements (with the general exception of vanadium) are used to colour ceramic glazes.

The effects obtained depend on the oxide composition of the base glaze, the amount of colouring oxide present, the thickness and uniformity of application and the atmosphere condition in the kiln during firing. Craft potters often take advantage of the different effects which can be obtained by deliberately varying the oxidising, neutral or reducing atmosphere in the kiln.

The *oxides of iron* are perhaps the most widely employed in glazes. Iron is usually introduced in the form of red ferric oxide, Fe_2O_3, but black ferrous oxide, FeO, magnetite, Fe_3O_4, hydrated ferric oxide (ochre) and iron-containing clays also find application. Crocus martis is ferric oxide, Fe_2O_3, produced from pyrites, and imparts bright colours to glazes. Dissolution of iron oxides from the body into a glaze can give interesting effects.

In oxidising kiln conditions the presence of iron gives a pale yellow, honey or brown coloration to a glaze. At levels of greater than 10% (Fe_2O_3) red crystalline (aventurine) effects are produced. In a reducing atmosphere a pale blue grey ($\frac{1}{2}$–1% oxide), green or blue (2–5%) or black (8–10%) colour is obtained.

Tenmoku glazes, originally associated with the Sung Dynasty, are high firing raw glazes containing 8–10% iron oxide and are usually fired at least partially under reducing conditions.

Black *cobalt oxide*, Co_3O_4, is the most powerful colorant for glazes, giving at levels of below 1%, a strong blue colour. The cobalt easily dissolves in the glassy glaze matrix and enters the structure.

The very strong colouring effect of cobalt oxide requires that it be well dispersed to avoid the appearance of specks. This can best be achieved by milling together frit, clay and cobalt oxide. Cobalt carbonate has a lower Co content than the oxide and in addition a finer particle size; consequently it is often preferred to the oxide to alleviate the specking problem. Cobalt compounds have a fluxing action in a glaze.

Chromium oxide, Cr_2O_3, imparts a green coloration to some glazes whilst in other compositions red, yellow, pink or brown colours can be produced. This oxide is highly refractory and it is not normally employed at levels greater than 2–3%. Additions of up to $1\frac{1}{2}$% results in complete dissolution of the oxide, but higher additions give opaque greens.

Although the colour normally associated with chromium is green, a red crystalline effect can be achieved by the addition of chromium oxide to a high lead, low alumina glaze maturing at 800–900°C, but the poor durability of such a glaze precludes its industrial use. The addition of zinc oxide in the presence of chromium oxide in a glaze results in a brown colour developing; yet if such a glaze has a high lead content, a yellow colour results.

In glazes containing tin oxide, a high level of alkaline earth oxides and silica, a pink can develop with a $\frac{1}{2}$–1$\frac{1}{2}$% chromium oxide addition. At temperatures above approximately 1100°C, glazes containing chromium oxide tend to lose this oxide to the kiln atmosphere through volatilisation. This can react with tin (opaque) glazes resulting in a pink coloration or with glazes containing zinc oxide to give a brown coloration. This effect is known as "chrome flashing".

The addition of *nickel oxide*, NiO, to a glaze can produce a wide range of hues, the colour being dependent on the composition of the glaze. The oxide is refractory and additions are usually limited to less than 5%. The usual colour which develops is brown but green and grey glazes can also be obtained. With high zinc glazes, dark blue colours can be achieved. Pinks and mauves can be produced when barium carbonate is present in the glaze. Due to its relatively unpredictable behaviour nickel oxide is rarely used as the sole colorant, it being normal to employ it in combination with other oxides.

Manganese dioxide, MnO_2 (or manganese carbonate) added to many glazes gives a range of browns, but red, purple and black colours can also be produced depending upon the basicity of the chosen glaze composition. Purplish colours develop on firing high alkali glazes containing manganese.

A well-known type of brown transparent glaze known as Rockingham results from the addition of a combination of manganese and iron oxides to a clear glaze. With appropriate additions of oxides of cobalt, iron and manganese, black colours can be produced.

Varied colours can be obtained by the use of *copper oxide*, CuO, in glazes, but due to the volatility of the oxide, its use is limited to where temperatures of approximately 1100°C are not exceeded during glost firing. Also, the use of copper oxide (or carbonate) in lead glazes can generate excessive lead release from the fired ware particularly with contents of copper greater than 1$\frac{1}{2}$% (CuO). Even from leadless glazes, soluble copper can be extracted by acidic juices which then acquire a bitter taste.

Used at levels of up to 3%, transparent green or turquoise glazes are produced when the addition is to a clear glaze and the firing conditions are neutral or oxidising. The turquoise colour is obtained by a 1–2% CuO addition to alkaline glazes, although crazing is usually a problem with such compositions. Lead glazes containing copper oxide fire to a green colour, particularly in the presence of zinc or alkaline earth oxides.

At higher levels of addition such that the glaze is saturated with copper oxide, metallic "pewter" effects are achieved.[8]

Under reducing atmosphere firing conditions, the presence of copper in a glaze can result in rich red colours being obtained.[9] Different shades of red develop depending upon the glaze composition and some of these effects have been given characteristic names (e.g. rouge flambé, sang de boeuf). Only minor amounts of copper (II) oxide (or carbonate), within the range 0.3–2.0%, are necessary as a base for these effects and a small quantity of iron oxide is usually present to take up preferentially any oxygen during the cooling cycle. A small amount of tin oxide is also added to stabilise the colour, which arises from the reduction of the copper (II) state to copper (I) and colloidal metallic copper. These effects are most easily achieved in glazes containing sodium, potassium and calcium fluxes. Light reduction starts at temperatures between 600°C and 850°C (depending on whether firing is to glost earthenware or stoneware temperatures), continuing through to peak temperatures and during the initial part of the cooling cycle.

Including silicon carbide in a batch is a means of inducing a reducing atmosphere during the firing of a copper-containing glaze. Only a small quantity of silicon carbide is necessary; otherwise an undesired rough surface texture may result

Vanadium pentoxide, V_2O_5, which can give brown or yellow colours, is rarely used. It is employed to a limited extent by craft potters, although large amounts have to be included in glazes even to achieve a medium strength yellow.

Minerals which can be utilised in glaze colouring include ilmenite, rutile, basalt, slate and various slags produced in the metal smelting industries.

Ilmenite and rutile are raw materials for the manufacture of white titanium dioxide pigments which are extensively used in the paints and plastics industries. The naturally-occurring minerals are, however, far from white and in their unrefined form are widely employed to give special effects.

Ilmenite is essentially ferrous titanate, $FeTiO_3$, and gives spotted and break-up effects when incorporated in glazes. Rutile in its pure form is white, but the mined mineral contains iron, chromium and vanadium oxides. Mottle and break-up effects can result from its use, although its colouring power is less than that of ilmenite. Ground basalt imparts a brown colour to glazes because of its iron content, but substantial quantities are often required.

Due to the unpredictable colour properties of the oxide colorants, stable manufactured inorganic pigments are used in the tableware, tile and sanitaryware industries. These are produced almost exclusively by the high-temperture calcination of oxides or materials giving oxides on heating. The raw materials usually include compounds of the first row transition elements—vanadium, chromium, manganese, iron, cobalt and nickel, all of which are capable of producing coloured compounds.

The products of calcination include newly formed crystalline phases (e.g. spinels), inclusion pigments, in which a colouring chromophore is entrapped in a crystal lattice (e.g. zircon), and solid solutions of chromophores. Compared with some simple oxide colorants, manufactured pigments are largely insoluble in a glassy glaze matrix. Cobalt silicate pigments are a notable exception, producing a vivid transparent blue coloration in an appropriate glaze. However, optimum results with many of the other pigments are achieved by employing glazes with specific compositional characteristics to reduce the already limited solubility even further. These compositional features are related to the nature of the particular stains themselves.

The commonly-employed stain types are presented in Appendix V in which preferred glaze compositions for certain pigments are indicated together with other features.

7.7.2 Cadmium Sulphide/Selenide Yellow, Orange and Red Glazes

The pigments containing cadmium sulphoselenide enclosed in a zircon lattice which are now available can be employed with no restriction on their host glaze composition. Such glazes can be fired to peak temperatures of 1300°C. However, cadmium glaze bases to which simple cadmium sulphide and selenide pigments are added have to be specially designed to give the required bright yellow, orange and red colours. These bright hues cannot be achieved by the use of oxide pigments yet they have always been popular.

The incorporation of cadmium sulphide in an appropriate glaze produces a brilliant opaque yellow, while increasing amounts of cadmium selenide in a cadmium sulphoselenide pigment results in orange, red and maroon colours. The yellow cadmium sulphide pigments usually also contain zinc sulphide and can be used in conjunction with the cadmium sulphoselenide pigments to produce intermediate tones. By themselves these cadmium pigments have insufficient stability to withstand the typical glost temperatures employed in the ceramic industries, but improved temperature stability can be obtained by using them in glazes with particular compositional characteristics. Nevertheless, the maximum temperature to which the most stable glazes containing these pigments can be fired is approximately 1100°C. For most cadmium glazes, recommended firing temperatures lie within the range 950–1050°C and their use is therefore restricted to appropriate tile, tableware and artware articles.

One method employed to improve the temperature stability of glazes containing cadmium sulphoselenide pigments is to incorporate cadmium oxide in the glaze formulation (e.g. in a frit).[10] Levels of CdO of 2–5% on the overall glaze dry weight are typically required to obtain the necessary stability in compositions containing 1.5–4% of cadmium sulphoselenide pigments. Alternatively, the sulphur and selenium can be incorporated into a frit with the cadmium being fritted into the same or another frit component. During the firing process, reaction between the cadmium present in one frit with the sulphur and selenium in another results in cadmium sulphoselenide being precipitated and the desired intense colour being obtained. Such glazes appear as white powders in the unfired state if no sulphoselenide pigments are present in the batch formulation.

The above types of cadmium glaze tend to have moderate to high thermal expansion due to the relatively high levels of alkali necessary to ensure good colour development. Consequently, crazing of this type of glaze is a problem.

An alternative method[11] of making cadmium glazes is claimed to overcome the problem associated with the high thermal expansion of the previously-mentioned glazes. Glazes of normal thermal expansion are produced with standard frits as the major glaze components (60–98% by weight) in conjunction with a "colour frit" in which the required cadmium sulphoselenide is formed and stabilised by the presence of materials such as zirconium silicate, aluminium oxide, tin oxide or titanium dioxide. Optionally present are lesser amounts of alkaline earth sulphates or antimony (V) compounds. The "colour frit" is produced by heating at temperatures above 850°C and generally in the range 1050–1250°C. The batch compositions also contain clays and/or other rheological agents.

No matter by what means cadmium sulphoselenide glazes are compounded, certain factors influence the ability to achieve good glost results. High levels of PbO (e.g. greater than 10–15%) can result in a blackening of the fired glaze as a result of the formation of lead selenide and lead sulphide. Commercial glazes, therefore, are either leadless or contain only limited amounts of PbO (5%). Excessive alkali content can result in attack on the colouring pigment and hence a deterioration in quality.

Care should be taken to ensure cadmium glazes are not contaminated in any way. Particles of metals such as iron, copper, zinc and aluminium in the glazes generate black spots and staining of the fired compositions.

Using cadmium glazes in conjunction with other ceramic pigments is usually avoided because of the risk of uncontrollable effects. However, a rich green glaze can be obtained by incorporating a cobalt blue stain in a cadmium sulphide yellow glaze.

Storage of some cadmium glazes in aqueous slip form can lead to the evolution of unpleasant fumes and only sufficient slip to meet immediate requirements should be made. Additionally, segregation of frit and pigment constituents can occur on standing.

Very thin applications of cadmium sulphoselenide glazes show discoloration of the fired film. Optimum results are obtained by applying a slightly greater than normal thickness.

Prior to firing, thorough drying of the glazed ware improves results after firing. In addition, ware should not be packed tightly in the kiln nor in the immediate vicinity of ware coated with other types of coloured glazes. If the latter precaution is not observed, vapours evolved by the other coloured glazes (e.g. copper glazes) can adversely affect the cadmium glaze. Conditions in the kiln should be oxidising and a relatively rapid firing cycle should be employed to avoid colour fading.

With the advent of the zircon encapsulated cadmium sulphoselenide pigments,[12] which have much improved temperature stability and do not require specific base glaze compositions, greater freedom has been given to the glaze formulator, although the pigments originally developed give pastel colours rather than the vivid tones of the simple cadmium sulphoselenides. The recent discovery of a method to produce zircon encapsulated cadmium pigments which can give glazes of intensity and vividness should result in their increased use.

7.8 TRANSPARENT OR OPAQUE GLAZES

Transparent glazes when fired are of uniform refractive index such that much of the light incident at the air/glaze interface is transmitted through the glaze before being reflected or scattered from or partially absorbed by the substrate. Opacity is the result of diffusion, reflection and refraction of incident light from particles or bubbles in the glaze.

This scattering occurs because the particles have a different refractive index (either higher or lower) from the surrounding glassy phase. The opacity of the fired glaze depends on the concentration of the opacifying phase within the glaze, the "particle" size of the opacifier, the refractive index difference between the glassy and opacifying phase, and the glaze thickness. The greater the refractive index difference and the closer the size of the individual opacifier particle is to the wavelength of light, then the greater will be the opacity produced. The dispersed phase does not have to be crystalline. Gas bubbles can contribute to opacity, as can the development of a different glassy phase. "Milkiness" develops in borosilicate melts when different high B_2O_3 phases separate.[13]

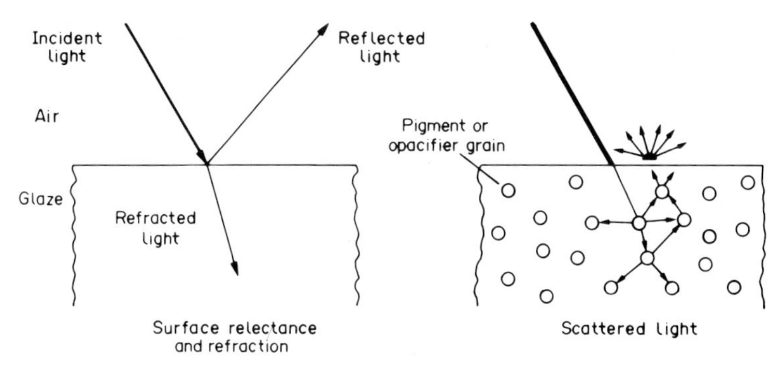

Fig. 7.1 Pigments refract light passing though them. In its passage through a glaze, light encounters pigment grains and is refracted, partially absorbed or scattered.
A ceramic pigment's ability to refract light is a function of the difference between its refractive index and that of the surrounding glass. The opacity of a pigmented system will depend also on the concentration of the pigment, its particle size, and the thickness of the glaze layer.

When it is necessary to hide an undesirable body colour and where aesthetic appeal is an important factor, opaque glazes are used.

If it is required that the colour of the substrate should show through the glaze, then an uncoloured transparent composition will be employed. Transparent coloured glazes will apparently modify the colour of a substrate. Transparent glazes have no crystals suspended in them which would interfere with their quality. If crystals form as the result of devitrification the surface can have areas of mattness. Devitrification is prevented by cooling the glaze rapidly through the temperature range 900–650°C.

The most widely used transparent glazes are colourless and applied to white firing bodies such as earthenware, vitreous hotelware, bone china and porcelain. Underglaze decoration is commonly employed with such glazes, the transparent coating allowing the decoration to show through and at the same time protecting it from physical and chemical attack.

There are different phases which can develop opaqueness in a glaze. All are not equally effective in their hiding power. Materials which can opacify glazes can be grouped according to type.

1. Crystals, either insoluble or not readily soluble in molten glaze, added to the mill and ground with other ingredients (titania, zircon or tin oxide).
2. Crystals which develop after controlled thermal treatment of the glaze (wollastonite).
3. Gas inclusions in the glaze (fluorine or air).
4. Liquid/liquid phase separation in the glass matrix.

Opacifying materials used in glaze, together with their refractive indices are given in Table 7.1

TABLE 7.1 REFRACTIVE INDICES

Zirconium silicate, zircon	1.96
Zirconium oxide, zirconia	2.35
Tin (Stannic) oxide	2.04
Titanium dioxide–anatase	2.52
Titanium dioxide—rutile	2.76
Cerium oxide	2.30
Antimony oxide	2.25
Calcium phosphate	1.63
Calcium fluoride	1.43
Air	1.00029

The clear glassy phase of a leadless glaze has a refractive index within the range 1.5–1.6. For a lead glaze the value lies in the range 1.6–1.8.

Tin oxide has long been employed as a glaze opacifier, but its high cost during the 1914–18 war resulted in the search for a cheapter alternative and led to the use in the U.S.A. firstly of zirconia and later of zircon. In 1950, ground zircons were commercially produced in the U.K. for the first time[15] and in the following years replaced the much more costly tin oxide. Today, tin oxide usage in glazes is generally restricted to specialist low-temperature and craft pottery glazes.

If tin oxide is added as a mill addition, then due to its low level of solubility in normal glaze systems, often no more than 5% is necessary for the development of full opacification. This opacifier gives the glaze a blue–white appearance.

Potash Felspar	65
Fine ground limestone	11
Fine ground silica	11
Zinc oxide	9
Tin oxide	4

Recipe for tin opacified glaze.

Zirconium silicate has a higher solubility in glazes than does tin oxide and therefore greater amounts of zircon opacifier are required to match the opacity of a tin oxide glaze. However, simply replacing tin oxide by zircon in a glaze formulated to use tin oxide does not usually result in the level of opacity being matched even if an increased amount of zircon opacifier is used. Due to the higher solubility of zircon in glazes, compositions containing this opacifier have to be specifically designed to reduce its solubility to as low a level as possible, commensurate with the appearance of the fired glaze. High contents of zircon modify the glossiness of the glaze.

High SiO_2 and Al_2O_3 levels in zircon glazes help to increase opacity, but both lead to increased glaze viscosity. A molecular SiO_2:Al_2O_3 ratio of 10:1 with the SiO_2 level being 2.5–3.5 moles, dependent on the required maturing temperature, usually forms the basis of satisfactory zircon–opaque glazes. The alkali metal oxides Li_2O, Na_2O and K_2O assist in obtaining the fluidity necessary to achieve a smooth fired finish but lead to increased zircon solution in the glaze and therefore should be used judiciously (0.3 mole total).

Zinc oxide is widely employed in zircon glazes because it increases whiteness and opacity, although in high-temperature opaque sanitaryware glazes it acts as a vigorous flux and can lower glaze viscosity markedly. A 0.1–0.2 mole zinc oxide content is typical of these glazes. Calcium oxide is a cheap flux used in most types of glaze, but in zircon opaques it impairs opacity, as does B_2O_3, which, however, is a valuable flux and does help to reduce thermal expansion. Lead oxide is a powerful flux but tends to give zircon glazes a cream tone, indeed even lead-free zircon glazes have a more yellow colour compared with tin–opaque glazes.

In zircon glazes the opacity may be achieved by incorporating zircon in a finely-divided form, either as a mill addition, as a frit component or by a combination of these two methods. When incorporated into a frit, a coarser grade of zircon can be employed than when it is added as a mill addition. This is due to the fact that during the firing of a raw glaze dissolution of the zircon does not occur as readily and the opacity of the fired glaze is dependent to a large extent on the particle size of the zircon used. The finer the zircon employed, the more opaque is the glaze. Optimum opacification occurs with a particle size for the occluding phase of 0.3 μm. Particles of this size are best obtained as a recrystallised phase. In the manufacture of zircon-containing frits, some solution of zircon in the glassy matrix does occur. Subsequently, on firing a glaze containing the frit, zircon in a very finely divided form is thrown out of solution with great benefit to the opacity. In fully fritted glazes the precipitation of crystals can be encouraged by the inclusion of a small amount of finely ground zircon in the mill.

Zircon with a particle diameter of less than 50 μm is a suitable grain size for use in fritting, while for use as a glaze mill addition, opacifier grade zircon with 95% of particles (by weight) less than 7 μm in diameter must be employed to attain satisfactory results.

Many zircon-opacified frits when rapidly quenched in water after fritting are transparent, indicating the high degree of solubility of zircon in borosilicate matrices. However, on reheating a sample of this frit, crystallisation of firstly zirconia and then of zircon occurs and opacity develops. With a particular high "zircon" content frit, which after quenching was reheated, formation of zirconia was evident at temperatures of 800°C, but at higher temperatures this converted to zircon.[16] Depending on the composition, zirconia exists in both monoclinic and tetragonal forms.[17]

Zircon glazes, whether fritted or raw, are more refractory than tin opaque glazes and higher fritting temperatures or longer frit kiln residence times are generally necessary to produce satisfactory frits than is the case with other frits. The complete incorporation of the relatively coarse zircon used in frit manufacture is essential if batches of glaze with uniform opacity are to be obtained. Shade variations arise from other factors.[18] Even small variations in the quantity of opacifier must be eliminated. Processing of glazes must be consistent; therefore, mill loads and milling times must be closely controlled. Opacification is influenced by the grain size and fluid slip gravity of the glaze base. The glaze should have a wide maturing range.

Depending on the composition of a fritted glaze, a zircon content of 5% may be required before any marked opacification develops in the fired glaze, but a content of about 10% zircon in a fritted opaque glaze is typical. Some dissolution of the fine zircon used in opaque raw glazes also occurs, the amount generally increasing as the particle size of the zircon is reduced, but on cooling, some recrystallisation of this dissolved zircon occurs. A major use of milled zircon products is in raw sanitaryware glazes where zircon contents of 9–15% are typically employed in the standard white opaque compositions. Although generally supplied as a dry product, the milled material has also been marketed in the slip form in which state it is claimed to have a greater opacifying power than the equivalent amount of the dry product of the same particle size. This is due to the better state of dispersion.

TABLE 7.2

Typical Zircon Opaque Frit Composition

Molecular formula (M.F.)		
0.443 Na_2O		4.210 SiO_2
0.048 K_2O	0.387 Al_2O_3	1.072 B_2O_3
0.220 CaO		0.389 ZrO_2
0.289 Zno		
Percentage (w/w)		
5.67 Na_2O	8.14 Al_2O_3	
0.93 K_2O	4.86 ZnO	
2.55 CaO	15.40 B_2O_3	
9.89 ZrO_2	52.18 SiO_2	

Zirconia, while lower in cost than tin oxide, is more expensive than zircon and it is therefore not widely employed as a glaze opacifier although it was used in the U.S.A. prior to the ready availability of milled zircon products. Zirconia in typical glaze compositions forms, during the firing process, zircon[19] and therefore any possible advantage accruing from the higher refractive index of zirconia compared with zircon is lost.

Comparison of zircon and tin oxide reveals technical advantages and disadvantges from the use of zircon. Zircon glazes do not suffer from the "chrome flashing" problem associated with white tin opaques which are coloured by the volatile chromium compounds present in chromium stains in glazes fired nearby. On the other hand, incorrectly formulated or overfired zircon opaque glazes suffer from a segregation fault which results in the appearance of flocs of opacifier, caused by an undue glaze solvent action on the zircon. When a proportion of this solvated zircon recrystallises, it does so in relatively large aggregates, causing loss of opacity. In coloured zircon glazes segregation can result from dissolution or volatilisation of certain "stain" or ceramic pigment ingredients. Soluble salts in a ceramic pigment also lead to segregation problems.

Although titanium dioxide, usually added as rutile, is widely used for producing opacity in glass enamels and in vitreous enamels, it does not find wide application as the opacifying phase in glazes which mature at higher temperatures. This is because the easy formation of large crystals of rutile gives the glaze a yellow coloration.

However, white opaque glazes in which titania is incorporated into a suitable frit composition can be obtained. In these glazes crystallisation of sphene, a calcium titanosilicate, $CaTiSiO_5$, occurs during firing of the glaze.[20,21] Good opacification can be achieved by crystallisation of sphene with a TiO_2 content of approximately 6–7% in the glaze. A typical glaze, preferably fully fritted and having 10% china clay as the rheology control agent, has the following composition (Table 7.3).

TABLE 7.3

	Wt%	M.F.
SiO_2	58.80	5.3
TiO_2	6.92	0.4
Al_2O_3	6.78	0.3
B_2O_3	12.84	0.85
CaO	6.86	0.55
BaO	2.38	0.08
Na_2O	4.03	0.33
K_2O	0.98	0.04

The development of unwanted colour tones in this glaze was influenced by the composition of the raw materials.[22] An acceptable iron oxide content was found to be 0.02% Fe_2O_3 (wt%). Good control of fritting and the glost firing is necessary to ensure consistent results from batch to batch. Addition of pigments to this type of base glaze can lead to reaction between the stain and residual titanium dioxide, resulting in colour tones which in many instances are different from those which would be obtained in a zircon-based opaque glaze. Sphene-opacified glazes give particularly intense oranges with Ti-Cr-Sb stains. Similar effects can also be produced by replacing the lime content of frits destined for sphene-opacified glazes with magnesium to produce compositions from which magnesium dititanate crystallises out. Both sphene and magnesium dititanate white opaque glazes tend to develop a slight yellow coloration, although this is not as marked as when titania is the sole opacifying phase in the glaze. In the range 920–1040°C, increasing peak temperatures further decreases whiteness.

Calcium phosphate is a rarely used opacifier because of the ease with which it blisters in glaze. It is useful only in low-temperature glazes where the interaction between glaze and opacifier is negligible.

Antimony oxide, although used in vitreous enamels, is not employed as an opacifier for glazes due to its instability at the high temperatures of a normal glost fire.

Finely dispersed bubbles of gas, such as fluorine, can produce a degree of opacity, but this method is not a viable alternative to solid phase opacification except in glasses. The development of a high level of opacity through gaseous inclusion is difficult to control in the thin layers of a normal glaze layer. Fluorine is found as a component of another phase, such as calcium fluoride. Opacification by air is most apparent in underfired glaze and it is rarely employed in normal practice.

Matt, vellum and crystalline glazes are also partially or fully opaque, but these special types of composition will be discussed in detail later.

7.9 SPECIAL EFFECT GLAZES

Matt, vellum, crystalline, glossy-opaque and many coloured glazes depend for their effect on the presence of crystalline phases with different refractive indices from that of the glossy glass matrix.

However, the use of the term crystalline glaze is usually meant to indicate that crystalline particles which are visible to the naked eye are present in the fired glaze. Matt glazes are ones in which the individual crystals producing the matt effect are not visible to the unaided eye. Vellum glazes exhibit a smooth semi-matt finish.

7.9.1 Matt Glazes

These glazes are produced either by inducing crystals of very fine particle size to develop during the firing cycle or by adding to a base glaze an excess of crystalline material such that much of it remains unaltered after completing the firing cycle. The crystallinity gives the matt appearance to the glaze.

The two commonest types of matt glaze are lime matts and zinc matts, the former produced by the presence of wollastonite crystals, $CaSiO_3$, in the fired glaze, and the latter by the presence of willemite, Zn_2SiO_4.

Lime matts can be produced by the inclusion of high amounts of ground calcite (limestone) or whiting in a glaze formulation in which silica is the major glass-former. Alternatively, an addition of ground wollastonite itself or use of a high CaO containing frit can give the same effect.

Zinc matts are formulated from a base glaze, having an addition of a matting agent which is itself made by prior calcination of zinc oxide and china clay in ratios of 1:1 to 8:1. Calcination increases the bulk density of the initial starting materials and the presence of alumina assists the formation of minute zinc silicate crystals rather than larger ones which might otherwise develop. In addition, with those matting mixtures based on high china clay content, the calcination destroys the plastic characteristics of the clay which could otherwise give wet–dry shrinkage problems after application of the glaze.

Where mattness is produced by means of crystallisation from the molten glaze during the cooling cycle, control of this cycle is clearly important. Too rapid cooling can result in a glossy appearance being retained because insufficient time is available for devitrification to take place fully. In certain circumstances, the characteristics of the body affect the nature of the matt effects produced because of modification of the glaze through glaze–body reaction. The fired thickness of the glaze can affect the development of crystalline phases within the glaze; the thicker the glaze is, the more reproducible will be the matt effect.

Matt surfaces can also be produced by suitably high mill additions of alumina or bone ash. Underfiring a glaze produces a similar (matt) appearance, but because such glazes have not properly matured, their other properties will not be at an optimum level and the use of such methods is not good practice.

Matt glazes exhibit a greater degree of surface roughness than glossy glazes and depending on the type of composition they can be either more, but are generally less, chemically and physically durable than the latter type of glaze. They are also more susceptible to metal marking and staining, both of which can be difficult to remove.

7.9.2 Satin or Vellum Glazes

These glazes have a smooth finish, but no individual crystals can be distinguished by eye. Such surfaces can be obtained by the addition of up to 18% stannic oxide or zinc oxide and up to 4% titanium dioxide[23] to high lead glazes and firing at 950–1000°C. At higher temperatures attainment of these effects is more difficult because of the variable degrees of solution of the individual oxides.

The base glaze is highly fluxed by lead oxide and boric oxide and satin glazes are very fluid at glost temperatures. They are usually restricted in firing to horizontal surfaces.

For the correct development of the vellum surface the zinc oxide should remain isolated in the glaze, but active enough to bond with neighbouring atoms of silica and titania. If this bonding does not occur and the zinc oxide acts only as a flux thereby entering the vitreous structure all crystallinty is destroyed and the glaze is glossy.

Although vellum glazes tend to be less susceptible to metal marking, the chemical durability and resistance to lead release is often poor, thus restricting their use in non-foodware items.

7.9.3 Crystalline Glazes

Glazes classed as crystalline are those in which individual crystals can be distinguished on the glaze surface and in the glassy matrix. Growth of the crystals is closely controlled by the nature of the kiln cycle,[24] consequently, the manufacture of uniformly decorated products is often difficult for many of the required types of surface texture. The only crystals of direct value are those which grow mainly in one plane, because the thickness of the glaze layer limits the size of an isometric crystal to a maximum size of about 0.5 mm.

At the glost temperature the matrix is wholly vitreous. Cooling the randomly stuctured glass enables the appropriate bonds between ions to develop and form crystalline structures. Given time these precipitate from the glass. In multiple oxide glaze the rate of growth is influenced by the presence of trace impurities acting as catalysts. Solution of the substrate alters the rate of crystallisation in direct proportion to the degree of solution.

Slow cooling in the region of 800°C is essential if crystal growth is to proceed and attainment of an adequate number of crystals can be difficult when fast-firing schedules are employed. True crystalline glazes are required to form saturated solutions at the peak kiln temperature followed by crystallisation during the (slow) cooling period. The crystals, usually needle- or lath-shaped, grow in a base glaze chosen for its low viscosity at the devitrification temperature. They are initiated by "seed" or nuclei of crystalline material. Titania is often the seed.[25]

The "rutile break-up" is a crystalline glaze effect which develops around coarse grains of the mineral rutile. The effect is still widely employed in the tile industry. The "break-up" depends upon the formation of lead titanate, $PbTiO_3$, with, optionally in addition, zinc titanate. Such glazes[26] have low SiO_2 and Al_2O_3 contents to maintain low glaze viscosity in the crystallisation range. The PbO content is usually high and rutile is present at 8–10 weight%. Crystallisation of the titanates takes place around titania crystals and therefore the initial particle size and shape of the rutile grain markedly influences the nature of the effect produced. As the crystals develop they absorb iron compounds from the surrounding glaze; consequently, the size and colour density of the crystals is dependent on the source (and therefore the chemical composition) of the mineral rutile sand. Ilmenite is used as an alternative seeding agent.

Depending upon the TiO_2, content of the glaze, either individual crystals which appear to be floating on a glossy background are observed, or, at higher TiO_2, levels, a more overall matt crystalline effect is evident.[27]

Matt glazes produced through the crystallisation of willemite (zinc silicate) have already been mentioned. Similar compositions, but with lower Al_2O_3 and SiO_2 contents and usually some TiO_2, have characteristics which allow crystallisation to proceed so that large individual crystals of willemite form readily. Claims are made that crystals on the glaze can be seeded by dropping dry crystals on the molten surface.

Aventurines are glazes in which small crystals are present in the fired glassy matrix and these exhibit a marked sparkling effect. Such glazes are produced by an addition of 10–15% of iron oxide (Fe_2O_3) to a transparent very solvent base glaze, although additions of 30% of iron oxide are known. Dissolution of the oxide occurs during the glost heating cycle, but on cooling crystallisation takes place to give the characteristic aventurine effect. Both high- and low-temperature glazes can be formulated to reproduce the crystalline effect, but lead glazes, low in alumina, are preferred. Other colouring oxides can be included to modify the effect of, or replace, the iron oxide. Chromium, nickel and manganese oxides can produce aventurine effects, but the optimum quantity is difficult to define. Too little will not give any crystals. An excessive amount of colouring metal oxide can result in the fired glaze having coarse crystals at the surface to give a metallic surface finish. Usually a medium-to-heavy application of glaze is found to be necessary, because thin coatings cannot give a sandy finish. The speed of cooling influences the size and quantity of crystals.

After the crystals have grown to a certain size and cooled to room temperature, reheating to the devitrification temperature will not encourage further growth. The cycle from the glost temperature must be repeated.

7.9.4 Textured Tear Glazes

This type of glaze is characterised by exhibiting, after firing, areas of matt surface which are surrounded by glossy channels. The major crystalline phase present in these glazes is zircon.

As with the "rutile break-up" crystalline glaze, these compositions are characterised by having low SiO_2 contents (0.7–1.3 moles typically) and high flux contents.[28] Zircon contents of 20% by weight are common. This may be added totally as a mill addition or split between a frit and a mill addition. Fritted zircon tends to produce a finer tear than does milled zircon. The tear effect can be developed in both lead and leadless formulations, the latter type generally requiring high B_2O_3 levels to enable the desired effect to be produced on firing. These glazes can be formulated for firing at peak temperatures in the range 900–1200°C. The low-silica/high-flux nature of the glazes produces in them a corrosive effect on the body and different results develop on substrates of different porosity. A more matt finish is often observed with more porous underlying bodies.

As with most types of textured/crystalline glaze, application technique has a major influence on the final glaze appearance. With spraying, the tear (channel) becomes greater as the application weight increases. Three-dimensional effects can be obtained using splatter application methods.

7.9.5 Crackle Glazes

Crazing is almost always considered as a fault in industrially-manufactured pottery, but on occasions, and deliberately, ware is produced with a hairlike pattern of fine crazes over the glazed surface. These glazes are described as crackle glazes. Such effects should not be considered as desirable on any foodware items because traces of foodstuffs will tend to collect within the craze lines and this could lead to subsequent contamination of food by bacteria. The use of crackle glazes is therefore more common in the craft sector where products are often of a purely decorative nature.

The crazing develops by encouraging the glaze to contract more than the underlying body during the cooling cycle, consequently glazes containing higher than normal amounts of soda and potash are required. In raw glazes, high levels of nepheline syenite or felspars can give the required high thermal expansion characteristics. However, for glazes maturing at a low temperature the high amounts of alkali must be introduced at least in part by the use of high-expansion alkali-containing frits.

The craze pattern in crackle glazes can be affected by the rate of cooling in the kiln. Rapid cooling tends to produce a finer craze pattern than that which develops with slower cooling. Although the crazing effect may be observed when the ware is in the kiln, the pattern may become clearly apparent only after its removal from the kiln.

7.10 SCRATCH AND SLIP RESISTANT GLAZES

Ceramic glazes are employed in a number of applications in which they are exposed to hostile treatment. Floor tiles are subjected to harsh treatment and they must resist scratching and exhibit slip resistance to prevent personal injury. For glazed floor tiles, these properties must be built into the glaze. For tableware it is necessary that cutlery does not scratch the glaze and conversely that the glaze does not scratch the metal such that metal-marking of the glazed ware occurs.

Floor tile glazes can be made scratch-resistant by ensuring that the exposed glaze surface has a high degree of hardness. This can be achieved by incorporating coarse particles of high hardness materials such as alumina and zircon into the glaze such that after firing these particles protrude out of the glaze in a textured surface finish. Alternatively, a smooth matt surface finish may be obtained with the crystalline phases having a high degree of hardness so that scratching is difficult. Glazes of the latter type in which crystals of zircon, Ca–Ti–O phases, willemite, gahnite ($ZnAl_2O_4$) or rutile are produced have good abrasion resistance.[29] However, the presence of willemite or gahnite can result in poor chemical resistance. The greater the amount of the abrasive crystalline phase in the fired glaze, then the better will be the overall abrasion resistance characteristics.

Glossy and smooth surface finishes generally have poor slip resistance and can often be scratched relatively easily. Tableware glazes need to be resistant to scratching by cutlery and this property is influenced by the nature of the glaze. When scratching does occur, there is a shatter pattern emanating into the glaze and it is the magnitude of this shatter pattern which is a measure of the scratching fault. For those transparent glazes usual on tableware, leadless glazes more readily give an extensive shatter pattern than high lead glazes. Low lead glaze compositions occupy an intermediate position.

Another undesirable feature which can impair the surface appearance of tableware plates and saucers are scuff marks resulting from the glazed articles impacting and sliding over each other when being stacked. In this case, leadless glazes are more resistant to the fault than high-lead glazes. Low-lead glazes again hold an intermediate position.

7.11 GLAZES FOR TABLEWARE AND COOKINGWARE

The major types of ceramic body used for producing tableware and cookingware articles are:

Hard paste or felspathic porcelain.

Bone and vitreous china.

Earthenware.
Vitreous alumina.
Vitreous earthenware.
Stoneware.
Cordierite (for oven-to-tableware).

Whatever the nature of these products, their glazes exhibit appropriate properties, for example acceptable metal release or low coefficient of thermal expansion. Many of the required properties need to be considered when designing glazes for them.[30]

The properties glazed tableware and cookingware need to exhibit are:

1. Low metal release characteristics so as to comply with legislative measures introduced by many countries. In the U.K. it is necessary for ware to comply with Statutory Instrument 1975, No. 1241, which requires ware to conform to BS 4860, Parts 1 and 2. (as at 1985)
2. Resistance to alkaline detergents employed in cleaning soiled ware is a property of particular importance for ware used by hotels and catering establishments. These products can be exposed to repetitive washing actions in the presence of quite corrosive alkaline detergents in dishwashing machines. Increasing use of such machines in the home means that ware is subjected to a harsher washing action than when hand washed and, therefore, better alkaline detergent resistance needs to be built in to domestic ware than was previously necessary.
3. Resistance to leaching of components from the glaze and/or body by foodstuffs, in particular drinks which might affect the taste of the food or drink.
4. Resistance to metal marking, scratching, scuffing, tea and coffee staining during normal use.

Other properties of the ware which might be of importance depending upon the use to which the product is to be or is likely to be put include:

1. High degree of thermal shock-resistance for freezer-to-cooker ware and oven-to-tableware.
2. Complete or almost complete transparency to microwaves so that the ware can be used in microwave ovens.

7.11.1 Metal Release from Glazed Surfaces of Tableware and Cookingware

Regulations have been introduced by many industrialised countries since the late 1960s as consumer protection measures. These stipulate the maximum amounts of lead and cadmium which may be leached from glazed ceramic ware by an acetic acid solution in contact with it under stipulated test conditions (see Chapter 9). Those two elements only are specified in most standards (e.g. BS 4860 and 5103 and ISO 7080,6486), although other metals such as antimony and barium are included in some national regulations. In these alternative regulations the type of acid leachant and the test temperature could also be different.

For tableware, the test is usually carried out at ambient temperature.

Some countries, for example the United Kingdom, require a separate test to be performed on cookingware and this includes a period of heating of the acid solution in contact with the ware (BS 4860, Part 2).

Metal release limits are usually expressed in milligrams per square decimetre of exposed surface (mg dm^{-2}), milligrams per litre (mg l^{-1}) or parts per million (ppm), the latter two being almost equivalent in the case of 4% acetic acid, the usual extractant. Limits in mg dm^{-2}, when stipulated, are usually for flatware (e.g. plates) when the area in square decimetres is the area of the horizontal plane encompassing the circumference of the article. For hollow-ware articles, the limits are almost exclusively expressed in milligrams per litre (or parts per million).

Limits set for cadmium are always lower than those specified for lead, reflecting the greater toxicity of the former element. Most regulations stipulate cadmium limits one-tenth those of lead, the justification for this differentiation being somewhat arbitrary. Lead limits as specified in various regulations range from 2 to 20 mg l^{-1} (ppm) depending on the type/filling volume of the ware.

Glazes for tableware or cookingware may be either lead-containing or leadless. Any onglaze or inglaze decoration might also contain lead, this generally being the rule for onglaze decoration. Similarly, cadmium might or might not be present in the glaze or in any decoration.

For colourless transparent lead-containing glazes, lead release has rarely been a problem with ware on which these compositions alone are applied. Any elevated lead release results observed from ware glazed with these lead-containing compositions are usually as a consequence of the presence of onglaze decoration on areas of the articles which are exposed to the leachant acetic acid during the test. However, gross underfiring of the glaze can result in high metal release. It is different for coloured glazes. Colouring oxides influence the release of heavy metal ions such as lead and zinc.[31]

Onglaze enamels contain glassy fluxes which allow this decoration to mature at temperatures appreciably below those employed to mature the glaze. These low-melting fluxes, whether they are lead-containing or leadless, therefore, have lower chemical durability than the glazes on which they are applied. With only a relatively minor change in composition the chemical durability of low melting fluxes can be markedly affected by a degree depending upon the change in composition. During the enamel fire, reaction between the enamel flux and glaze occurs, the degree to which such reaction proceeds being influenced by the firing temperature and by the viscosities of the glass phases at enamel firing temperatures. The lower the viscosity of the glaze at these temperatures, then the greater will be the degree of interaction. The nature (lead or leadless) of the glass compositions of the enamel and glaze can influence the lead released on exposure to acid following such interaction.

Where both the glaze and enamel flux contain lead, as is the case with onglaze decorated china and much earthenware, firing to higher temperatures in the enamel kiln usually results in lower lead release. This is due to the increased interaction between the enamel and glaze improving the durability of the enamel flux/glaze composite because the excellent acid durability of the more thickly applied glaze has an increasingly predominant effect. At the temperatures employed in the onglaze decoration of china and earthenware, (750–840°C), low-solubility lead glazes have softened to an appreciable degree and interaction is more significant than is the case with a hard porcelain glaze. With the very soft enamel lead-containing fluxes which found widespread use prior to the introduction of metal release regulations, very marked changes (e.g. 50- and 100-fold) in metal release can be observed after modifying the peak enamel temperature by 50°C. Onglaze colours prepared from the improved durability lead fluxes, which have been introduced in recent years, do not exhibit these effects in such a marked manner because they have very low metal release properties throughout their recommended firing ranges. Improvement is possible with some of these new compositions if the firing temperature is increased within the recommended range.

Using transparent low-solubility lead glazes in conjunction with leadless onglaze colours might be thought to offer a means of complying with metal release regulations when both components in isolation have excellent, very low or zero lead release characteristics. However, due to the interaction between the lead glaze and leadless colour during the decorating fire, excessive lead release can occur. To achieve the maturing characteristics required in the leadless enamel colour, a leadless frit containing relatively high amounts of alkalis and B_2O_3 must be employed and interaction between flux and lead glaze produces phases of relatively poor durability following diffusion of PbO and/or alkalis and B_2O_3 into the different glass structures. With such a combination of enamel and glaze, lead release increases with increase in the peak enamel temperature reached within the firing range of the enamel and then decreases as the firing temperature is further increased. This reduction is the result of the poor durability (high lead release) phase being further modified by the more durable glaze. It follows that with the inclusion or addition of say alumina or zirconia in the enamel, phase separation would be modified and a detrimental change in the observed durability would not occur.

Only by using leadless onglaze colours on a leadless glaze and firing in a lead-fire kiln atmosphere can freedom from the restriction of lead release be guaranteed.

Where cadmium is contained in the enamel composition, as either a frit component or as the pigment cadmium sulphoselenide, release of the cation cadmium follows this same pattern.

Use of inglaze decoration usually results in metal release from the ware obtained being similar to that observed from the glazed surface where no such decoration is used. In the in-glaze decoration process, the applied colour sinks well into the glaze and, therefore, has little effect on the durability of the glass surface exposed to the acid test liquid.

In addition to modifying the metal release characteristics of decorated tableware or cookingware by altering the peak enamel temperature, the composition of the glaze itself can be altered such that it has a different viscosity at a particular enamel firing temperature. For low-solubility lead glazes a reduction in the CaO or MgO content produces a lower viscosity at temperatures within the range 750–840°C. Using such a glaze, with lead-containing onglaze colours and a particular peak enamel temperature, lower lead release from the decorated ware usually results. With leadless onglaze colours the opposite effects might be observed.

However, when modifying the viscosity characteristics of a glaze at enamel firing temperatures, care must be taken not to impair the overall acid-durability of the glaze itself. An excessive increase in alkali or B_2O_3 content could reduce the durability of a lead glaze such that lead release (from ware decorated with approved lead-containing enamel colours) actually increases as a result of the inferior durability of the glaze.

7.12 EARTHENWARE GLAZES

Glazes applied to earthenware bodies, which are characterised by having appreciable porosity after firing (5–8% water absorption), can differ markedly in their compositional characteristics. Transparent glazes are almost exclusively used on hard paste procelain, vitreous and bone china because of the translucency of these substrates. Glazes for earthenware bodies can be transparent or opaque, coloured or uncoloured, matt, semi-matt, crystalline or glossy in appearance. This diversity in glaze effect is as a consequence of the body itself being opaque. Earthenware glazes invariably contain frit, but both once- and twice-firing techniques are employed in the manufacture of the ware. Flatware is produced by a twice-firing process, but hollow-ware items are often made by once-firing methods.[32]

Where twice-firing techniques are employed, biscuit firing is normally at peak firing temperatures of approximately 1150°C with the subsequent glost fire being at temperatures of 1050–1100°C. However, some manufacturers, usually producing hollow-ware articles (lamp bases, vases), employ an easy biscuit (1020°C) and hard glost (1150°C) process. For hollow-ware, particularly mugs produced by once-firing, peak temperatures of approximately 1150°C are necessary to give the required fired properties.

In the U.K. transparent glazes used on earthenware tableware are most often low-solubility (fritted) lead compositions. Transparent leadless glazes are employed on hollow-ware articles and although these compositions are also usually fritted,[33] raw leadless glazes can be used in once-firing processes. Coloured glaze, for earthenware is either lead or leadless.

The lead-containing glazes employed in the U.K. are normally based on a combination of lead bisilicate and one or more transparent leadless borosilicate frits with clays and optionally other materials such as felspars, calcium carbonate and zinc oxide. Lead borosilicate frits are used to a lesser degree than in glazes for bone or vitreous china.

Leadless fritted earthenware glazes usually contain higher proportions of frit than lead-containing compositions. Zircon–opaque fritted glazes are widely employed on mugs produced by once-firing methods. Leadless fritted glazes are used to a lesser extent on such ware.

For earthenware tableware (tea and coffee sets) both underglaze and to a generally lesser extent onglaze decorating techniques are employed. With underglaze decoration, the fired glaze must be transparent or only semi-matt to allow the underlying print to show through. Alternatively, the print might be hydrophobic and "resist" the glaze during the dipping operation and give a pattern in reverse relief. Care, however, must be taken to ensure that constituents of the glaze do not adversely affect the desired decorative effect. In particular, the amount of ZnO in a glaze has to be restricted if underlying colours include certain pink, brown and green pigments. As little as 1% ZnO in a glaze can modify the colour imparted by chrome oxide, Cr_2O_3, calcium chromium silicate (Victoria Green) and zinc-free brown underglaze colours.

Where lead-containing onglaze colours are applied to fired low-solubility glazes, reduction in the level of lead and cadmium release from the final article by modification of the glaze composition is not as readily achieved as in the case with vitreous bodies. For the latter types of body, we have noted, lowering of the viscosity of the glaze at enamel firing temperature by compositional modification can result in reduced metal release, but applying the same technique to an earthenware glaze results in an increased susceptibility to "spit-out" (Chapter 12). Indeed, for earthenware which is to be subjected to an onglaze decorating fire, a *high* glaze viscosity is desirable to prevent the spit-out fault. Consequently it is necessary to balance the requirement for high viscosity at enamel temperature for spit-out resistance with that for low viscosity for low metal release.

7.13 GLAZES FOR BONE AND VITREOUS CHINA

Bone china was discovered in England by Josiah Spode and from the late eighteenth century has been produced almost exclusively in the U.K. Lead-

TABLE 7.4 BONE CHINA AND EARTHENWARE GLAZES

Glaze /Type	ref./Substrate		Composition										Firing	Thermal Expansion
			SiO_2	B_2O_3	Al_2O_3	Na_2O	K_2O	CaO	BaO	PbO	ZnO	ZrO_2	Temperature (°C)	Coefficient ($\times 10^{-6}$ °C^{-1})
A	Bone china	wt.%	49.18	8.24	10.47	5.25	1.86	6.15	—	18.90	—	—	1050	6.42
	Transparent	M.F.	2.74	0.40	0.34	0.28	0.04	0.37	—	0.28	—	—		
B	Bone china	wt.%	55.44	11.12	10.94	5.75	1.04	7.45	1.00	7.26	—	—	1080	6.00
	Transparent	M.F.	3.35	0.58	0.39	0.34	0.04	0.48	0.02	0.12	—	—		
C	Bone china	wt.%	56.70	6.56	11.12	4.19	1.50	4.98	—	14.95	—	—	1080	6.30
	Transparent	M.F.	3.94	0.39	0.46	0.28	0.07	0.28	—	0.28	—	—		
D	Earthenware	wt.%	67.27	10.85	8.19	11.65	0.19	1.85	—	—	—	—	960	5.90
	Transparent	M.F.	5.02	0.70	0.36	0.84	0.01	0.01	—	—	—	—		
E	Transparent	wt.%	52.17	7.07	9.97	2.95	2.67	7.07	—	18.09	—	—	1020	5.90
	Transparent	M.F.	3.07	0.36	0.35	0.17	0.10	0.45	—	0.29	—	—		
F	Earthenware	wt.%	53.60	6.75	9.75	2.92	0.53	7.55	—	18.90	—	—	1080	5.30
	Transparent	M.F.	3.28	0.36	0.35	0.17	0.02	0.50	—	0.31	—	—		
G	Earthenware	wt.%	58.92	15.82	11.70	4.49	—	2.45	6.62	—	—	—	1080	7.00
	Transparent	M.F.	6.16	1.43	0.00	0.46	—	0.27	0.27	—	—	—		
H	Earthenware	wt.%	61.25	14.22	11.33	6.08	—	7.12	—	—	—	—	1100	6.50
	Transparent	M.F.	4.53	0.91	0.49	0.44	—	0.56	—	—	—	—		
I	Earthenware	wt.%	57.54	13.26	12.00	5.35	—	6.72	5.13	—	—	—	1100	6.50
	Transparent	M.F.	4.00	0.80	0.49	0.36	—	0.50	0.14	—	—	—		
J	Earthenware	wt.%	56.10	7.45	9.00	5.93	—	6.35	—	—	6.48	8.67	1120	4.70
	Opaque	M.F.	3.24	0.37	0.31	0.33	—	0.39	—	—	0.28	0.24		

containing transparent glazes which give the product its characteristically brilliant surface have always been used. Until recently, decoration was exclusively by the use of onglaze colour and although the glazes currently employed are extremely durable from the point of view of lead release, using onglaze colours (whether leadless or lead-containing) can result in problems of compliance with lead release regulations in a limited number of cases. The durability of onglaze decoration to alkaline detergents is relatively poor and with the increasing use of dishwashers and corrosive detergents there has been a trend towards decoration underglaze for bone china. As a result the product has both negligible metal release and improved resistance to the effects of alkaline solutions. With felspathic china, similar trends in the use of glazes and colours have occurred.

Biscuit firing of these bodies to a temperature of approximately 1220–1230°C usually precedes the glazing operation in which peak firing temperatures of 1060–1120°C are reached. In contrast to the high temperature porcelain fire, oxidising conditions are employed throughout the bone china glost fire. Glazes are combinations of lead borosilicate, lead bisilicate and leadless borax frits with additions of clays and optionally other minerals such as nepheline syenite and felspar. Once-firing china hollow-ware with raw leadless glazes is a practical alternative.

Excessive metal (lead and cadmium) release can result from the use of onglaze colours, even when the fired glaze by itself is of acceptable durability. Even if both the transparent glaze and the enamel colour contain lead, the glaze will have the greater chemical durability as a result of its more siliceous composition. Any reaction between the glassy components of the glaze and enamel during the decorating fire will tend to result in an improvement in durability of the composite compared with that of the enamel component in isolation. Increased reaction therefore generally results in lower metal release. Reaction can be increased by employing a glaze of lower viscosity at the peak enamel decorating temperature and this is achieved by selecting a glaze with a CaO content of $\simeq$1% which is low compared with levels of 7–9 wt.% CaO often found in china glaze compositions.

7.14 LOW EXPANSION GLAZES

Although the coefficients of thermal expansion of the common tableware bodies span a reasonably wide range (approximately 5×10^{-6}°C^{-1} for porcelain to 8.5×10^{-6}°C^{-1} for bone china) the glost temperatures employed with these bodies are such that the production of glazes with the required thermal expansions is not difficult.

However, when glazed articles are required to withstand very sudden changes in temperature, as for example with freezer-to-oven ware, not only are low thermal expansion ceramic bodies necessary but the glazes used on them must be low enough in expansion coefficient to remain in compression at all times.[34]

Bodies with a very low expansion can be formulated within the system, $Li_2O–Al_2O_3–SiO_2$ from which β-spodumene and/or β-eucryptite phases crystallise during firing. Very low thermal expansion substrates can also be formulated from compositions which develop crystals of cordierite $Mg_2Al_4Si_5O_{18}$ on firing. Such bodies have coefficients of thermal expansion in the range $1.4–4.0\times10^{-6}{}^\circ C^{-1}$ depending on the overall composition. Cordierite bodies have a number of technical applications where excellent thermal shock resistance and electrical insulation is required. These applications often do not require a glaze but in the case of freezer-to-ovenware, a glaze coating on the product is a necessity. Problems have been experienced in formulating very low coefficient glazes, so self-glazing cordierite compositions have been developed[35] but these cannot generally be relied upon to produce consistent results from successive firings.

Glazes with the required thermal expansion characteristics can be formulated from these same two systems,

$$Li_2O–Al_2O_3–SiO_2 \text{ and } MgO–Al_2O_3–SiO_2.$$

Both raw and fritted glaze compositions which develop crystals of β-spodumene and/or β-eucryptite have been described and these glazes often exhibit a highly crystalline surface as a result of the differences in refractive index between the crystalline and glassy phases. Some formulations which have been described include agents such as ZrO_2, TiO_2 or P_2O_5 which influence the rate of nucleation and/or crystallisation.

Raw glazes include, as the source of Li_2O, a lithium aluminium silicate mineral such as petalite. Amounts of other alkali metal oxides should not exceed 2% by weight in total, otherwise the development of higher expansion phases leads to crazing.

A raw, petalite-based glaze with a coefficient of linear thermal expansion of $2.4\times10^{-6}{}^\circ C^{-1}$ has been described.[36] This glaze had the batch constitution as indicated below:

Silica sand	43% by weight
Alkaline felspar	10%
Petalite	30%
Alumina (calcined)	7%
Talc	5%
Dolomite	5%

It was applied to either a green or biscuit body of initial composition: talc 40%, alumina 18%, clay 25%, petalite 7%, felspar 7%, and ZnO 3%. Glost firing at 1250–1300°C produced a white, semi-matt glaze.

Fritted low thermal expansion glazes within the Li_2O–Al_2O_3–SiO_2 system have been extensively studied. One of the first glazes reported[37] was compounded from a frit of composition (wt%):

SiO_2	50.4%	P_2O_5	2.6%	K_2O	1.0%
Al_2O_3	29.2%	TiO_2	2.6%	B_2O_3	2.8%
Li_2O	5.9%	Na_2O	1.0%	PbO	2.8%
ZrO_2	1.7%				

to which was added 5% china clay in the mill.

This glaze, with a coefficient of thermal expansion of 2.4×10^{-6}°C^{-1}, gave a matt yellow-brown finish. The recommended firing schedule for the glaze followed glass–ceramic practice in that nucleation and crystal growth stages were stipulated, the TiO_2, ZrO_2 and P_2O_5 acting as nucleating agents and/or crystallisation rate promoters. The presence of TiO_2 was the cause of the coloration.

White firing fritted compositions which also give surfaces ranging from matt to relatively high gloss have also been described.[38] These are recommended for use on very low expansion bodies designed for cookingware. The glaze compositions were either fully fritted or contained only a small proportion of clay or other minerals and were found within the compositional range (wt%) Li_2O 4–23%, MgO 0–6%, Al_2O_3 17–40%, SiO_2 36–74%, ZrO_2 0–5% and flux 5–20%, where B_2O_3, K_2O, F, PbO, Na_2O, CaO, SrO, ZnO and BaO were considered as flux components. One composition from within this range was compounded by grinding (in water) 100 parts of frit of the following composition (wt%):

SiO_2	54%	MgO	2%
Al_2O_3	27%	B_2O_3	4%
Li_2O	9%	K_2O	4%

with four parts of bentonite. Applying this glaze to a low-expansion biscuit fired body, firing it to 1100°C with a 2-hour soak and cooling over 10 hours resulted in a craze-free glaze of moderate gloss. Other compositions described varied from fully matt to glossy, opaque to transparent and had thermal expansions within the range -8 to $+31\times10^{-7}$°C^{-1} (50–350°C). None of the glazes contained nucleating agents, although some contained ZrO_2 as a crystal growth rate promoter. This ingredient was found to influence the glossiness of the surface giving reduced gloss with compositions of low expansion and added gloss to glazes with relatively high expansion.

Other crystallisable glaze compositions within the Li_2O–Al_2O_3–SiO_2 system have also been described.[39] For these the presence of one or more nucleating agents (ZrO_2, SnO_2 or TiO_2) was stipulated and the substrate on which the glazes could be applied was either a green or crystallised glass–ceramic of very low thermal expansion. The glazes described were fully fritted compositions having expansions as low as $-73.5\times10^{-7}\,^{\circ}C^{-1}$ and high Li_2O contents (16–19.5%). A frit composition which had a thermal expansion coefficient of $+8.7\times10^{-7}\,^{\circ}C^{-1}$ when fired to 1100°C on a specified cycle had the following composition (wt%): SiO_2 51.1, Al_2O_3 21.0, Li_2O 19.2, TiO_2 3.2, PbO 5.5.

Fritted glazes in which crystals of cordierite are produced have also been described.[40] These glazes have oxide compositions in which the following components are present (wt%):

SiO_2	66–75%	B_2O_3	2.5–7%
Al_2O_3	13–22%	Na_2O+K_2O	0.4–2.0%
MgO	3–8%		

Small amounts of other oxides might account for up to 5% of the glaze composition according to the claims of the patent. Glazes within this compositional range are stated to be transparent since the cordierite crystal and the glassy matrix have similar refractive indices. Typically the glazes have thermal expansion coefficients of $2.8\times10^{-6}\,^{\circ}C^{-1}$ and they are recommended for freezer-to-ovenware made from a cordierite body.

The refractory nature of these compositions does however require very high fritting temperatures (1550°C). The glazes, produced preferably by wet grinding the frit with 4% bentonite, were found to require firing temperatures of 1200–1350°C. The best results were obtained by applying the glaze to a cordierite body, previously biscuit fired at a temperature appreciably below the glost temperature. Once-firing techniques were also found to give satisfactory results.

Glass–ceramic substrates have a wide range of thermal expansions and mechanical strengths. Applying a compressive glaze to a high expansion substrate can increase the modulus of rupture fivefold.[41] Low-expansion glazes fire well on some glass–ceramics and they develop a strong intermediate layer of interlocking crystals. Glaze compositions are found in the system:

$$Na_2O–CaO–PbO–B_2O_3–Al_2O_3–SiO_2$$ [42]

with coefficients of expansion in the range $40–80\times10^{-7}\,^{\circ}C^{-1}$. As an example of a low expansion glaze a composition of (wt%)

SiO_2	60.0%	K_2O	1.4%
Al_2O_3	6.9%	CaO	4.5%
B_2O_3	11.6%	PbO	15.3%
Na_2O	0.3%		

will fire to a glossy surface at 1100°C with a 1-hour soak. The major portion of the composition is fritted and ground with china clay as a suspending agent.

7.15 SANITARYWARE GLAZES

The production of sanitaryware is today carried out by a once-firing process, firing temperatures being within the range 1150–1270°C. For domestic sanitaryware, the body is a vitreous china composition made from a felspathic flux, quartz, china and ball clays. Fireclay bodies are also employed, particularly for heavy duty applications with the application of an engobe to hide the undesirable buff body colour.

From the hygiene viewpoint, manufacture of a vitreous product has advantages, but it is still necessary to ensure that the ware can withstand the effects of relatively corrosive disinfectants and cleansing powders. In addition, the product should not be susceptible to the effects of staining materials. These properties must be built into the glaze and British Standard BS 3402 describes chemical tests for glazed vitreous sanitaryware.

As well as white opaque glaze, a wide range of coloured glazes is employed, particularly in the industrialised countries. The glazes are generally completely raw compositions, although a very small proportion of frit might be included to improve gloss.

Tin oxide was once widely used as the opacifier in sanitaryware glazes, but because of its very high price it has been almost completely replaced by zircon. Zircon based stains (e.g. vanadium blue, praseodymium yellow, iron coral and cadmium "red") are the major pigment constituents of most coloured glazes due to their excellent temperature stability. Zirconia–vanadium yellow, chrome tin pink, cobalt silicate blue and iron chrome brown stains are also used. To assist cleaning, glazes which fire to give a glossy finish are usually employed although certain manufacturers include semi-matt glazes within their ranges.

Not all ware produced by the once-fire process is of an acceptable quality. However, due to the high costs already incurred it is often economic to further process this ware to give articles of the required quality for sale. Most of the substandard ware from the first fire can be economically reclaimed by a further application of glaze and refiring so as to hide blemishes from the initial glost fire. The nature of the refire glazes and the necessary refiring glaze cycles differ in continental Europe and in the United Kingdom.

In mainland Europe, the glaze employed for recovery of substandard ware is exactly the same as that used in the initial once-fire process and consequently to refire any piece, a peak temperature equating to the first fire is chosen. In Britain leadless fritted (zircon) opaque compositions are employed as refire glazes; consequently, lower peak temperatures of between 1120–1150°C can be, and are, employed. These glazes are preferably free of B_2O_3 to eliminate faults arising from volatilisation of this component.

The speeds of both the once-fire and refire cycles are restricted both by the large size of the articles being fired and by the presence of crystalline silica (quartz) in the body. Crystalline silica produces a volume change at the α-β transition temperature and this necessitates the use of slow rates of temperature change in the vicinity of the transition point. Exposing the fired body to rapid temperature changes causes dunting.

Once-fire cycle times are (at the present time) typically 14–20 hours. Refire cycles, no matter what type of glaze is employed, are 20–24 hours long. The latter cycle is slower because dunting is a danger on both the heating and cooling cycle, whereas with a once-firing process dunting is a hazard only during cooling, after the body has vitrified.

Interest in the faster firing of sanitaryware is great and some reduction in firing times can be expected as new kilns, capable of giving an even heat distribution over a setting, are introduced.

Appreciably faster cycles will require reformulation of the bodies presently employed, but this is unlikely to result in major changes in the compositions of the glazes because the restricting factor is likely to remain the size of the article and the properties of the body. The chosen firing cycle will develop full gloss and remove bubble from the glaze.

In designing a glaze for the once-fire process it is essential that the composition does not seal over before evolution of gases from the body has been completed, otherwise pinholing or blistering in the glaze will occur. The glaze composition is controlled so that the glaze–body interface seals within the range 1090–1130°C. These raw glazes normally contain felspars or nepheline syenite as the major flux, but lithium minerals (e.g. petalite) are also used to a limited extent. The glazes usually contain, in addition to the zircon opacifier, CaO, which may be introduced as the carbonate or as wollastonite ($CaSiO_3$), ZnO, Al_2O_3, introduced as china clay or "Molochite" (calcined china clay) and SiO_2 as ground quartz. Barium carbonate, talc, dolomite and strontium carbonate are also possible constituents of sanitaryware glaze formulations.

Wollastonite[43] used in place of ground limestone or whiting as the source of CaO can reduce the incidence of the faults of pinholing and eggshell finish as a result of the consequent reduction in the amount of gas evolved. In combination with felspars, wollastonite forms an eutectic at 710°C and can result in ready dissolution of the other glaze ingredients in compositions containing these minerals. When whiting or limestone is the source of CaO, higher temperatures are necessary to initiate the dissolution processes. However, to equal the viscosity of a satisfactory whiting glaze at glost temperature, it is usually necessary to reduce the Al_2O_3 and SiO_2 contents of a wollastonite version of the glaze. If wollastonite is the major source of CaO in the glaze, then additions of 16–22 wt% are typical. Such glazes have a wider firing range as well as better gloss than glazes containing whiting.

Partial replacement of the clay content of sanitaryware glazes by calcined clay (e.g. "Molochite"TM) can also improve fired glaze quality by reducing the incidence of "pinholing".

Once-fire sanitaryware glaze bases—glaze without the opacifier—generally fall within the following molecular composition range:

$Li_2O+Na_2O+K_2O$	0.1–0.3	Al_2O_3	0.3–0.4
$MgO+SrO+BaO$	0–0.3	SiO_2	2.75–3.5
CaO	0.3–0.55		
ZnO	0.05–0.30		

Zircon contents are usually 9–13% of the dry glaze batch weight. Typical compositions are as follows:

Examples 1 and 2[44]

Yellow raw sanitary glazes using whiting and wollastonite as alternative sources of CaO.

	Eg1	Eg2
Felspar	27.0	29.4
Ground quartz	29.4	19.4
Wollastonite	—	24.4
Whiting	19.0	—
Kaolin	10.4	11.3
Zinc oxide	2.2	2.4
Zirconium silicate	12.3	13.4
Zircon praseodymium yellow stain	3.0	3.0

Example 3[45]

White opaque glaze for vitreous china sanitaryware.
The recommended firing cycle for this composition involves heating to 1200°C in 12 hours and soaking for 2 hours at that temperature prior to cooling.

Batch		Molecular formula (ignoring zircon)	
Potash felspar	26.3		
Silica	284.4	SiO_2	3.0
Calcium carbonate	3.2	Al_2O_3	0.3
Barium carbonate	5.9	CaO	0.325
Dolomite	11.4	BaO	0.1
Zinc oxide	4.7	MgO	0.2
Kaolin	10.3	ZnO	0.2
Zirconium silicate (Zircosil 5)	9.8	$(K,Na)_2O$	0.175

Sometimes this soaking period includes a short, gentle, cooling period to avoid the generation of new bubbles.

7.16 STONEWARE GLAZE

In the first making of stoneware by the Chinese, the pottery perhaps acquired a glaze by interaction of the body with fine ash from the wood fuel. Later glazes were prepared from finely ground felspathic minerals and consequently firing temperatures were high, around 1350°C. Some fusible clays were, and still are, used as a slip glaze. In the Middle Ages, the discovery was made that if common salt (sodium chloride, NaCl) was thrown into the kiln at peak temperature, the salt dissociated and the sodium combined with the clay to form a vitreous glaze.

For more practical glazes auxiliary fluxes are needed and alkaline earth additions markedly reduce the melting point of the base felspathoids and clays. Studio potters would add wood ash to the list of fluxes. Small quantities of frit can be included if the glaze is applied to a dry clay body. Boron and lead compounds are not used as fluxes because of their volatility at the high firing temperatures; nevertheless, stoneware glazes are easy to formulate. They can be applied to ware, in all its stages of making, by dipping or brushing, although it is normal to apply the glaze to the ware in the clay state.

When compounding the glaze a balance is maintained between the ratio of fluxes to refractory oxides, to prevent the glaze melting too soon and sealing in any gas still being evolved from the reactions in the body materials. At worst the glaze blisters and at best it is unsightly, whilst the body might bloat. Glazes with sealing points of below 1100°C are avoided.

Stoneware glazes become an integral part of the body and add to the strength of the ware. Quality of gloss is variable and surfaces range from the shiny to the highly textured. Reducing gases in kiln atmospheres not only develop characteristic effects in the glaze but they can also lower the melting point of some of the components in both body and glaze. (Ferrous oxide is a potent flux.) An imitation slip-glaze or engobe effect can be developed at the throwing stage if glaze slurry replaces the lubricating water normally used.

Many glazes reflect the subtle aesthetic properties of the stoneware body and possible textures range from the unctuous to the matt and from the transparent to the opaque. High firing temperatures mute all colours and the breadth of the palette is limited. Some of the useful metal oxide colorants are volatile at these high temperatures, and the available colours, commonly black, brown, yellow, green and blue, are developed mainly from iron oxide in its many configurations. Bright colours are rare.

7.16.1 Felspathic Glaze

Compositions of suitable glaze will usually contain up to 50% felspar and 20% fine silica, with the balance made up from clay and the alkaline earths.

	For 1300°C ("Hard")	For 1200–1250°C ("Soft")
Potash felspar	42	72
Fine ground calcium carbonate	18	13
China clay	25	7
Fine ground silica	15	8

The character of the glaze is determined by the quality of the raw materials (and their level of impurities) and by the way in which the glaze is used. Stoneware glazes are very scratch-resistant and usually they have good resistance to acid attack.

Engobes, sometimes called "dry glaze" can add to the range of effects on stoneware.

7.16.2 Wood Ash Glaze

Vegetable ashes contain varying amounts of all glass-making elements,[46,47] and in art glazes ash can be used as the fluxing agent, though usually only in glazes firing above 1150°C. The proportion of alkali to refractory oxides also varies from ash to ash, and the studio potter classes each ash according to its fluxing power—hard, medium or soft. Within this broad definition the compositional variation in other oxides produces a range of unique textures and colours in the glaze. Bark and twigs give the best ash; the older tree growths contain more silica. The ash of grass, nettles, reeds and shrubs are also used.

The wood is burnt in an open fire and when cold soaked in a large water-bath. Charcoal, floating as a scum, is removed on a coarse sieve. The bath is decanted to separate sandy grit and the suspension then passed through a finer sieve. After allowing the liquor to settle for several hours the supernatent liquor is removed. Fresh water is added until the suspension is free from soluble salts. The residue is dried and the yield of "alkaline fluxes" is about 50% by weight of the ash from the burning.

In a glaze the basic proportion is one part each of "ash", felspar and ball clay, but the substitution of a different plant ash will produce crazing where there was none, alter the colour and change the surface texture and no reliance can be placed on continuity of effect from successive batches of the same ash.

As examples, glazes for firing at 1250°C can be made according to the following formulations.

TABLE 7.5

	For clayware	For biscuitware
Potash felspar (300s mesh)	25	50
China clay	35	10
Fine ground silica	0–10	0–10
"Wood ash"	25–35	25–35
Fine ground calcium carbonate	5–15	5–15

TABLE 7.6 PERCENTAGE ASH CONTENT AND COMPOSITION FROM VARIOUS AIR-DRIED VEGETABLE MATERIALS

	% Total	% Composition of ash								
Substance	ash	SiO_2	CaO	MgO	Na_2O	K_2O	P_2O_5	SO_3	S	Cl
Wheat straw	4.26	66.2	6.1	2.5	2.8	11.5	5.4	2.8	3.8	—
Barley straw	4.39	53.8	7.5	2.5	4.6	21.2	4.3	3.6	2.9	—
Heather	3.61	35.2	18.8	8.3	5.3	13.3	5.0	4.4	—	2.2
Fern	5.89	6.1	14.1	7.6	4.6	42.8	9.7	5.1	—	10.2
Reeds	3.85	71.4	6.0	1.3	0.26	8.6	2.1	2.8	—	—
Sedge	6.95	31.4	5.3	4.2	7.3	33.2	6.7	3.3	—	5.6
Rush	4.56	11.0	9.4	6.3	6.6	36.6	6.3	8.8	—	14.2
Wheat (grain)	1.77	1.7	3.3	12.4	3.3	31.1	46.3	2.2	—	8.4
Beech (leaves)	3.05	33.8	44.9	5.9	0.7	5.2	4.7	3.6	—	0.3
Beechwood, trunk	0.55	5.4	56.4	10.9	3.6	16.4	5.4	1.8	—	—
Beech brushwood	1.23	9.8	48.0	10.6	2.4	13.8	12.2	0.8	—	—
Oak	0.51	2.0	72.5	3.9	3.9	9.5	5.8	2.0	—	—
Apple	1.10	2.7	70.9	5.5	1.9	11.8	4.5	2.7	—	—
Mulberry	1.37	3.6	57.0	5.8	6.6	36.6	6.3	8.8	—	14.2

7.17 GLAZES FOR ELECTRIC AND ELECTRONIC APPLICATIONS

For these applications, a wide range of ceramic bodies including felspathic porcelain, alumina, zircon porcelain, cordierite and steatite are employed and for certain uses glazing is necessary. In some cases the glazes must have particular electrical properties.

7.17.1 Conducting Glazes for High-Tension Insulators

For long distance electric power transmission, very high voltages are necessary to make the operation economic. A feature of the high-tension overhead power lines of the National Grid system in the U.K. is the large insulator strings. These insulators can be fabricated from a porcelain ceramic body.

Firing a raw transparent glaze onto such an insulator reduces the effects of any surface contamination to which the insulator is subsequently exposed. Although the glaze itself is an insulator, an electrically-conducting surface layer is produced through the action of water extracting alkali ions from the glaze structure. It has been reported[48] that an increase in relative humidity from 50% to 90% can reduce the resistance by a factor of 10,000. Even in dry conditions the voltage distribution over the surface of the insulator is not uniform and the situation is more variable in humid atmospheres. The presence of the conducting layer can lead to localised discharges or more seriously to complete flash-over and loss of the supply. Radio interference, particularly with pin-type modulators can also occur through discharge between conductor and insulator.

These problems have been largely overcome by the use of conducting glazes following work at the Central Electricity Generating Board in the early 1940s.[49] Having a conducting glaze on high-tension insulators stabilises the voltage distribution over insulator strings if the resistance of the glaze is comparable to that of the surface layer.

In the U.K. glazed procelain insulators are produced by once-firing to peak temperatures of approximately 1250°C; consequently, the glazes are the raw, leadless type. The first commercial conducting glazes of this type included Fe_2O_3 with additions of other oxides, for example Cr_2O_3, CoO, CuO, NiO and ZnO, so that conducting spinel phases were formed and distributed uniformly throughout the fired glaze. Prefired spinel pigments based on these oxides can also be employed to produce conducting glazes, 10–20% by volume of pigment being necessary to develop a continuous conducting network.

These compositions were a significant improvement on earlier glazes, but were susceptible to rapid corrosion, particularly around the electrodes. The corrosion phenomena have been investigated[50] and it has been shown that the rate of corrosion increases as the number of oxides constituting the spinel conducting phase increases. In the same investigation, a new type of conducting glaze containing an addition of 17–18% barium ferrite, $BaFe_{12}O_{19}$, and 5% ZnO to a base glaze was described. Such a glaze was found to exhibit a high degree of corrosion resistance.

Barium ferrite is itself non-conducting and therefore a conducting phase must develop during the firing of the glaze, in which the ZnO content plays a crucial role. It was initially thought that Fe_3O_4 was the conducting phase, but later work[51] indicated that conduction arose primarily as a result of the formation of zinc ferrite, $ZnFe_2O_4$, at a temperature of 900–1000°C during the glost fire, barium going into solution in the glassy matrix.

Conducting glazes can also be formulated by introducing oxide semiconductors into glazes.[52] Semiconductors form through the introduction of small quantities of impurity ions into an oxide lattice whose constituent metal ions are capable of existing in variable valency states. The doping of the host lattice can be developed by calcining an intimate mixture of the components.

Four-, five- and six-valent oxides introduced into an Fe_2O_3 lattice will produce a semiconductor crystal which in a glaze will give the required conductivity properties. However, these compositions have high temperature coefficients of resistivity, high thermal expansion and their corrosion resistance is not outstanding. Combination of Sb_2O_3 with SnO_2 by calcination at 1200–1300°C and subsequent addition of this to a suitable base glaze can give compositions which have the desired conductivity.[54] Variation in resistivity with SnO_2/Sb_2O_5 content has been found to be more uniform than is the case with simple Fe_2O_3-containing glazes. Glazes containing SnO_2/Sb_2O_5 semiconductors have low coefficients of thermal expansion (giving good mechanical strength), low temperature coefficient of resistivity and exceptionally good electrolytic corrosion resistance. Typically the semiconducting SnO_2/Sb_2O_5 phase contains 5% Sb_2O_5 by weight and comprises approximately 30% of the glaze batch.

Suitable formulations for different types of conducting glazes are as follows:

Barium ferrite glaze[53]

Base glaze A	83 wt%
Barium ferrite	17 wt%

The base glaze A has the composition:

	Seger mole formula	wt%
$(Na,K)_2O$	0.2	4.5
CaO	0.4	6.4
MgO	0.1	1.2
ZnO	0.3	7.0
Al_2O_3	0.4	11.7
SiO_2	4.0	69.2

(i) Recommended peak firing temperature: 1225°C.
(ii) Recommended soaking time: 2 hours.

Semiconducting glaze containing SnO_2/Sb_2O_5[55]

Base glaze B	65 wt %
SnO_2/Sb_2O_5 base (2.5 mole % Sb_2O_5)	35 wt %

Base glaze B has the composition:

	Seger mole formula	wt%
$(Na,K)_2O$	0.2	4.6
CaO	0.7	11.5
MgO	0.1	1.2
Al_2O_3	0.4	12.0
SiO_2	4.0	70.7

(i) Recommended peak firing temperature: 1225°C.
(ii) Recommended soaking time: 2 hours.

7.17.2 Glazes for the Electronic Industry

In electronics, two types of microcircuitary are employed—thin- and thick-film. Thin-film circuitry, which is the more precise but costlier technique, involves the application of the circuitry by vacuum techniques to the substrate. Photolithograph and etching methods are additionally employed. Subsequently, anodisation processes are used to produce dielectric components by conversion of deposited metal to metal oxide. Organic acids such as citric and oxalic acid can be employed as electrolytes in this process.

In thin-film circuitry the properties of the substrate are a major influence on performance. The substrate is required to act as an electrical insulator, a good heat conductor to rapidly dissipate heat away from the circuit, have good chemical resistance properties, mechnical strength and an even, defect-free surface finish. The high thermal conductivity, good mechanical strength and electrical insulating properties of alumina, together with the surface smoothness of glass which also acts as an insulator, suggested the use of glazed alumina as a suitable substrate for thin-film circuits.[56]

For this demanding application it is necessary that the fired glaze has a smooth, defect-free surface finish with no edge meniscus. Too much compression must not be built into the glaze, for this can lead to warpage of the wafer-thin alumina substrates. A thin glaze layer reduces the incidence of such problems and also means that greater advantage can be taken of the higher thermal conductivity of the substrate. A thin application of glaze (preferably approximately 25 μm) requires close control of the coating and firing process. Applying the glaze as a preformed transfer offers advantages over spraying; in particular the thickness can be more closely controlled.

Due to their low-viscosity characteristics at glost temperatures, fully-fritted lead borosilicate glazes give the best results. Electrical conduction within the fired glazes is minimised by maintaining the alkali content, particularly Li_2O and Na_2O, at a low level. If alkali is present it is usually K_2O only, the presence of the mobile alkali ions, Li^+ and Na^+, resulting in power ageing and corrosion of the applied tantalum film. The presence of large divalent ions, e.g. Pb^{2+}, Ba^{2+}, also helps to reduce ionic mobility. Bismuth oxide, Bi_2O_3, can advantageously be employed in the glaze composition as a flux to reduce the amount of PbO required and give improved chemical durability. Suitable glazes based on the $PbO-Bi_2O_3-BaO-Al_2O_3-SiO_2-B_2O_3$ system have been described.[56]

Interest in the use of glazed alumina for thin-film resistor and capacitor circuits has declined partly due to problems of ion-migration and cracking within the glaze and partly as a result of the major improvements made in the properties of alumina substrates. Initially these substrates were produced by a pressing process and were of a relatively poor quality, having surface porosity and a surface finish of 1.25 μm centre-line average (CLA). As a result of using better starting materials and processing techniques (tape casting), improvements in the properties of the fired alumina substrates have been achieved so that they can be used for the preparation of thin-film circuitry without the need for a glaze application to give better surface properties. Surface finishes down to 0.05 μm CLA can now be achieved.

"Thick-film" Circuits

Whereas thin film technology uses the glaze to act as a base for the circuits, thick-film technology uses the glaze as the circuit; consequently, the glazes are required to be resistors, conductors or dielectrics. The precise positioning of these circuit components is made possible by using the silk screen process and the application is direct to the ceramic. Print thickness plays a major part in reaching stability in trimmed thick-film resistors. The optimum fired thickness is 10 μm. Firing times to maturing temperatures in the range 600–850°C are short and the cycles are repeated for additional layers.

For conducting films the compositions are mixtures of metals or alloys and frits. The metals are selected from the noble metals. The frit is a lead borosilicate. For resistors the lead borosilicate frit is mixed with powdered palladium (which oxidises to form a vital component) and silver. For capacitors titanates are mixed with the frit.

7.18 STRUCTURAL CERAMICS

7.18.1 Heavy Clay Products

There are special problems to be faced when glazing natural clay bodies because of varying properties in the clay. Thermal expansion values differ according to the position or strata of the deposit. Contraction from dry to fired dimensions vary and the level of soluble salts can change between different sites within the same deposit; consequently, the need to adjust the properties of the glaze is ever present.

Climatic variations have an effect on glazed structural clay products and a frost-resistant body is selected. There must be a good bond between glaze and body. Rarely is there glaze peeling, but crazing is often encountered because some natural clay bodies have low thermal expansions. Some clay deposits are unsuitable for glazing, but they might be modified by the inclusion of pulverised fuel ash or expanded vermiculite, the presence of which will affect the requirements for a matching glaze.

Extruded or pressed bricks, with or without surface treatment, can be glazed either in a once-fire or a twice-fire process. Both lead and lead-free glazes firing within the range 950–1200°C are used.

Two families of glaze are available, the choice depending upon the firing temperature. Low firing glazes (often containing frit) are generally part of a twice-fire process in which the glost temperature is below the burning temperature of the already-fired brick or tile. High firing glazes are generally part of a once-fire process in which they are applied to the unfired brick. This is economically attractive, but the maintenance of a high-quality glaze surface is difficult. Prefired brick is the preferred substrate for glazing because the absence of body reactions (which disturb the glaze layer) allows rapid glost firing schedules.

Lead oxide is the main component of low firing glazes, not only because of its fluxing power but also because of its effect in lowering surface tension and improving mechanical properties. Glazes can be developed for the long firing schedules of some brick kilns. Low surface tension glazes easily coat porous brick bodies and cover imperfections in the brick surface. High fire glazes are usually lead-free. These often crawl on heavy clay products and to avoid such problems they are often used in conjunction with engobes.

Glaze suspensions having different rheologies will produce different effects on the same brick body which itself has received different surface treatments. Wire cut bricks have small protruding leaves which penetrate the glaze to give attractive texturing quite unlike the plain glaze on a pressed brick face or on an engobe. Transparent and coloured glazes can be fired on traditional sand-faced bricks.

To secure good adhesion and give good handling properties many additives are included in the glaze recipe. Organic hardeners are usually present to prevent flaking of the dried unfired glaze and to give it strength while setting in the kiln.

Tunnel kilns are the preferred firing method because of the cleanliness and ease of control of the furnace atmosphere. However, when suitably compounded, brick or tile glaze can be fired in conventional chamber kilns. There are restrictions on where the bricks can be placed in the chambers because the glazes require the atmosphere to be oxidising and free from fuel ash. The bricks should not be face set, nor side set, but "boxed" if possible to protect the glaze against the atmosphere. Broken tile is used to separate the bricks to prevent the glazed edges of adjacent bricks from fusing together.

Constructional heavy clay ware is subjected to weathering and the exposed glazed surfaces should therefore be acid-resistant.[57] Cyclic freezing and thawing tests should be carried out on the glazed unit.

TABLE 7.7 BRICK GLAZE COMPOSITIONS

	Na_2O	K_2O	CaO	ZnO	PbO	Al_2O_3	SiO_2	B_2O_3
Low firing glaze M.F. (Twice fire, 1000°C)	0.07	0.01	0.10	—	0.82	0.21	2.33	0.13
High firing glaze M.F. (Once fire, 1140°C)	0.23	0.21	0.37	0.192	—	0.57	4.63	0.08

7.18.2 Engobes

Where the fired colour of the ceramic body is dark and the colour of the glaze is to be light, or where it is necessary to equalise differences in stress between body and glaze, then an engobe is interposed as a compensating layer. Engobes are mixtures of clays, fluxes and fillers which coat the substrate with a permanent, opaque, layer. In some applications these layers are weatherproof and their compositions lie somewhere between a glaze and a clay. Engobes ranging from porous to semi-vitreous also provide a variety of surface treatments for heavy clay in their own right. When pigmented, engobes can be used to hide the colour of the (poor) clay base. (see Table 7.8)

TABLE 7.8 ENGOBES FOR BRICKS

White engobe	1150–1250°C	1000–1150°C
Potash felspar	20–30	15–25
Ball clay	15–20	10–15 (Ball clay and China clay combined)
China clay	15–20	
Silica	20–25	15–20
Whiting	5–15	0–5
Borax frit	0–5	15–30
Zirconium silicate	–	15–20

Engobes can be applied in one of several ways;[58] spraying, brushing, dipping or dry rolling. Where applied as a slip, the properties of the engobe are controlled by adjusting:

1. The ratio solids to liquid for shrinkage control.
2. The levels of flocculant or deflocculant for viscosity and thixotropy control.
3. The ratio of plastics to non-plastics for control of wet-to-dry and dry-to-fired shrinkage and "fit".
4. The level of organic hardener to control bond development between engobe and clay, the drying rate and abrasion resistance in the dry coating.

If the engobe has a high slip viscosity, then it can be applied to a wet extruded column of clay.

The preparation of the slip is similar to the compounding of a glaze with the added problem that the physical and chemical properties of the chosen ingredients have equal importance. To prepare an engobe composition which will fit all clays is impossible and each clay product must be treated individually. For high firing temperatures (1150–1250°C) the composition will be based on ball clay, felspar and silica with limestone (whiting) or zinc oxide. If the firing temperature is reduced, the complexity of the composition increases and talc and ground frit can be included.

As a guide, a basic slip made from 5 kg engobe powder dispersed in 2 kg water will cover 1000 bricks. Adjustment to the solids/water ratio will be made according to the chosen application technique. Adding an organic hardener to the slip enhances the bond between body and engobe, as well as producing a coating which can be handled without marking.

Any tendency for the engobe to peel or be rejected by the brick because of the presence of high levels of soluble salts in the clay can be overcome by the addition of up to 1% $BaCO_3$ to the body. This also eliminates a defect called scumming in the clay product and the engobe.

Application of Engobes to Bricks

There is a choice of application methods and either wet or dry techniques, sometimes with the addition of sand, have been used.

Wet application method. This can take the form of either dipping, spraying, brushing or rolling. The latter two methods have not found much favour in practical production, being difficult operations to automate and difficult to adjust for coating the brick headers.

The simple dipping of bricks in the engobe slip produces an uneven surface because of the high viscosity of the slip. This method is usually restricted to laboratory or small plant trials.

The most widely used system is spraying. There are several machines available which are fully automatic and ensure a thin uniform engobe coating on the brick surface. The machine sits astride the moving column of bricks. Two guns coat the face of the brick and side guns coat each of the two headers as the brick travels continuously through the spray booth. The minimum quantity of slip needed to mask the colour of the clay is sprayed onto the column so that when fired the engobe will also mask the natural colour of the brick. On textured brick surfaces a thin application covers only the high spots whilst hollows or troughs and cracks remain untouched. The appearance of the brick is a combination of engobe surface and background colour of the brick, yet giving an impression of overall colour when viewed from a short distance. Small chips on the bricks blend with the general effect. It will be noted that the equipment and technique for engobe spraying is similar to that used for glaze spraying.

Dry application. The engobe powder, either plain or stained, is sprinkled onto the brick clay column and then pressed, rolled or stippled to fix the colour on the surface. If necessary the brick surface has been softened by an application of a fine water spray. The result is a mottled surface similar to that of slip application on textured surfaces. This technique cannot be used to engobe the headers.

7.19 CONCRETE

Construction ceramics include concrete building blocks, and these can be glazed at low temperatures[59] without detracting from their strength. There is a maximum temperature below which glazing must be completed. Above 650°C the amount of cracking which develops in concrete reaches a point where the strength begins to fall[60] beyond the limit at which it can be regenerated. Being exposed to weather, the acid resistance of the glaze must be good and the difficulties of combining this factor with the correct thermal expansion and low maturing temperatures cannot be underrated.[61] These parameters point to the use of lead oxide as a major component of the glaze. The advantage of using the "mixed-base" effect can be seen in the examples in Table 7.9.

TABLE 7.9

Component	5-oxide System (wt%)	6-oxide System (wt%)
SiO_2	41.35	45.16
TiO_2	4.14	4.53
ZnO	2.91	1.59
Na_2O	7.76	7.27
PbO	43.84	39.12
Li_2O	—	2.33
Maturing temperature	620°C	600°C
% Loss in weight from grains[1]		
$H_2SO_4+HNO_3$	0.38	0.06
Acetic acid	0.08	0.00

1 see Section 9.3.1

The glazes are waterground with mill additions of colour or opacifier and organic hardeners. After adjusting fluid properties to suit the particular method of coating, the glazes can be applied either by spraying, dipping, brushing or even by paint roller. Surface texture is affected by the surface texture of the concrete and therefore any surface imperfections in the block are filled with an engobe composition of glaze and fine silica. After firing the concrete is recured by immersion in water.

7.19.1 Cement

A "glaze" on cement can be developed[62] by coating the article with an alkali metal silicate, drying the coating and then applying a second coating of an alkaline earth metal salt which converts the original layer to silica. The article is autoclaved to complete the reaction.

For both concrete and cement products (including sand-lime bricks) glossy or matt flame sprayed high lead glaze coatings are feasible. Provided natural or synthetic ceramic materials are added to the lead glass frits, crazing in the coating can be reduced to a minimum. The ideal method is a two-stage process—engobe and glaze.

Pinholing in the final coating is avoided by applying to the substrate a wash coat containing potassium silicate and aluminium oxide as an "engobe" before flame spraying a thin coating of glaze. Preheating the substrate by passing the surface several times under the flame also reduces the incidence of bubbling. The best results are obtained by first applying a thin flame-sprayed coating of a frit (Frit A in Table 7.10).

Over this base is sprayed a glaze made from a mixture of frit (Frit B in Table 7.10) and mill addition, the mill addition being used to control crazing. Materials which were found to be effective in this duty were alumina, fused silica and petalite.

TABLE 7.10[63]

	PbO	Li_2O	Al_2O_3	SiO_2	Na_2O	K_2O	B_2O_3	TiO_2	ZrO_2
Frit A	80	2.4	8.0	9.6	—	—	—	—	—
Frit B	53	4.0	—	23.0	4.0	2.0	12	1	1

7.20 PORCELAIN GLAZES

In the manufacture of hard-paste or felspathic porcelain tableware, the clay body is firstly subjected to a low-temperature biscuit fire (900–1000°C) prior to application of glaze and then firing to peak temperatures of 1300–1400°C so as to mature the glaze and body together. Reducing kiln conditions are employed at these high temperatures, this treatment giving the translucent product a blueish rather than a yellow tinge as a result of reduction of trace amounts of Fe(III) to Fe(II) present in the raw materials. The body is vitreous and therefore it has good physical strength which allows thin sections to be potted. As translucency is one of the major attractive features of hard-paste porcelain, glazes employed are normally colourless and transparent although occasionally they are semi-opaque. Due to the high temperatures of the glost fire, glazes are raw and leadless. Such glazes usually consist of felspar, quartz, kaolin (china clay), calcium carbonate and dolomite or other source of magnesia. They may also contain zinc oxide, lithia (introduced usually as petalite) and barium carbonate.

Hard porcelain glazes typically have silica contents of 79–83 mol% (approximately 75–75 wt%); consequently they have extremely good resistance to attack by acids and alkalis, and as a consequence glazed porcelain is used for laboratory ware (e.g. evaporating dishes). The high silica and low alkali content also results in procelain glazes having low coefficients of thermal expansion, typically $3–5 \times 10^{-6}$ °C^{-1}, and since the porcelain body has an even lower thermal expansion, porcelain tableware exhibits good thermal shock resistance. The highly siliceous nature of the glaze means porcelain tableware is extremely resistant to scratching by cutlery.

As glazes employed on hard porcelain are free of both lead and cadmium, toxic metal release is not a problem if the ware is undecorated. However, use of onglaze decorating techniques can result in the release of lead and cadmium ions if these elements are present in the decorating colours. A recent trend in the manufacture of hard porcelain table and cooking ware has been the use of inglaze decoration in which the ware is fired at 1200–1290°C on a fast (60–90 min) cycle at the expense of enamel firing at temperatures of 820–850°C. Inglaze decoration has excellent acid and alkali durability.

Onglaze decoration involves the use of lead-containing colours and results in inferior durability, yet the superior durability of inglaze-decorated ware is achieved when either lead or leadless colours are employed.

Typical hard porcelain glaze compositions are as follows[64] (Seger formula):

(a) 0.314 MgO   1.120 Al_2O_3 10.300 SiO_2
    0.140 CaO
    0.167 K_2O
    0.379 Na_2O

(b) 0.246 MgO   0.711 Al_2O_3 6.570 SiO_2
    0.550 CaO
    0.068 K_2O
    0.107 Na_2O
    0.006 Li_2O

The recommended peak firing temperature for both glazes is 1390°C.

7.21 TILE GLAZES

The range of glazes applied to ceramic tiles covers almost every type of composition (e.g. lead/leadless, fritted/raw, once-/twice-fired, opaque/transparent, matt/glossy).

This is due to the diverse characteristics of the different types of substrate composition used in various parts of the world where fired body properties can range from relatively high expansion to low expansion, from high porosity to vitreous and white to highly coloured.

Ceramic tile products can be split into two types, wall and floor tiles. The manner in which they are to be used, perhaps for internal or external wall cladding, influences the properties which have to be built into the body and glaze. Tiles for external use need to be resistant to the effect of frost which might, with an unsuitable substrate, lead to breakdown of the body. This happens with porous bodies when water absorbed in the pore system is subjected to freeze/thaw cycles with the body experiencing an expansive stress as the absorbed water freezes. For external cladding, vitreous tiles ensure frost-resistance, but for internal use porous tiles can be employed. Internal floor tiles are usually fully vitreous as these types of tile have better strength properties and will be resistant to any possible frost damage. Vitreous tiles can be fitted as internal wall covering, but they tend to be more expensive to produce than the lower temperature firing porous tiles.

Other properties required in glazed tiles are introduced by the glaze itself. This is the case with floor tiles which need to have slip- and scratch-resistance. All types of glazed tiles have to be resistant to staining and the effects of chemical cleaning agents.

The regular shape of ceramic tiles and the fact that they are glazed on one face only has resulted in production being highly mechanised with many processes fully automatic. This is true of glaze application and many different techniques are employed to glaze ceramic tiles. Simple waterfall and spraying methods can give either a complete or partial covering of the tile's (upper) surface, but many other techniques only give a partial covering of glaze on the tile. Special units are built into the glazing line to apply glazes by the desired methods. At its simplest, the glazing line might involve passage of the tile through a single waterfall unit; a more complex line could include a waterfall unit, one or more spraying booths plus a number of more specialised units (e.g. sponge mottler, screen-printer), with different glazes being applied at each station. Methods of glaze application relevant to tile production are discussed in more detail in Chapter 8.

Ceramic tile manufacture is a major area where fast-firing techniques are being employed on a large scale, the main reason for this lying in the flat shape of tiles and the fact that they are glazed on one side only. These features allow high setting densites in single-layer fast fire kilns. Presently, glost firing of biscuit-glazed tiles is being achieved on cycles of as little as 30 min duration whilst once-firing on cycles of 45–60 min is also being carried out. At these rapid rates, body reactions are different from those generated in longer firings. In addition there is reduced activity at the glaze–body interface and the extent of crystal formation within the glassy matrix will be changed—indeed some components will not have been fully incorporated into the structure. The consequently modified thermal expansion characteristics will affect glaze fit. Glazes for porous wall tiles are at least partially fritted to enable the compositions to mature without vitrification of the body taking place. For simple white or coloured opaque glazes, leadless formulations can be used but often a small amount of lead bisilicate is added to reduce the viscosity of the fired glaze to promote its flow. Opaque glazes usually have zircon as the opacifying phase and this material may be added as a mill addition, a frit component or a combination of the two methods. The use of high-lead glazes is widespread in order to obtain many of the patterned effects which can be achieved on tiles. Such glaze compositions may be the only type of glaze applied to the substrate (e.g. textured glaze, rutile break-up glazes) or be one of several glazes employed in a multi-application process.

7.22 PREPARING GLAZE FOR USE

The quality of the deposition of glaze on the substrate is directly proportional to the consistency of the slip. This usually increases as the slip density ("pint weight") increases; consecutive batches of the same glaze must have the same particle size and be prepared according to the same proportions of powder to water.

Each glaze will have its characteristic, optimum, particle size whose value depends on the type of glaze and the method of application. Normally glazes have size distributions such that 55–75% (wt) of particles are <0.01mm.
The fluidity of the slip is affected by any change in the particle size.

Viscosity is adjusted until the condition of the glaze is appropriate for the size, design and type of ware and the chosen method of application. In any stock of slip glaze the viscosity will be adjusted daily by the addition of dry glaze material, glaze slip with a higher fluid density, or electrolytes.

The optimum condition for glaze is the (undried) water-ground and fully dispersed state of a slip, ball milled from the raw ingredients, but there are many occasions when the supply of glaze is as dry powder and this must be reconstituted.

7.22.1 Reconstituting Glaze

By Hand Mixing

Prepared glaze in powder form or "raw" glaze mixes must be reconstituted as completely as possible, but the completeness of the original milled dispersion of the precurser raw materials is never equalled. Dry powder can be given a preliminary mix by hand or (for larger quantities) by mechanical stirrer, but subsequent sieving is necessary. Glaze powder is always added to water.

For small quantities, sieving is a satisfactory technique for glaze dispersion.

The dry glaze, mixed with a proportion of the water required, is pressed through a sieve with a stiff brush. After a first pass through a sieve with a mesh opening of 180 μm to 250 μm the glaze is allowed to "mature" for several hours. When the glaze is required it is sieved again through a 106 μm to 125 μm mesh. The addition of hardeners and electrolytes is made at this stage, followed by resieving. The fluid density is adjusted.

Base glaze can be modified by sieving with the addition of colouring agents (either fine ground or "specks") and during this treatment the slip is passed through a 75μm–125μm sieve.

By Milling

Repeat charges of the same glaze supplied in the dry state can be efficiently dispersed on a vibro-energy or ball mill running for 30 min only. Care is taken to avoid further reduction in particle size, limiting the action in the mill solely to one of breaking down agglomerates formed on drying the original glaze slip. Leadless glaze is dispersed with 40% water. Lead glaze is dispersed with 33% water. The fluid density is accurately adjusted after discharge from the mill.

From the mill the slip is passed through a 160 μm mesh to remove any residual agglomerates and reduce the risk of specking. Where large quantities of glaze are being reconstituted, use of a vibrating sieve is normal practice.

REFERENCES

1. Roberts, G. J., Salt, S., Roberts, W. and Franklin, C. E. L. *Trans. Brit. Ceram. Soc.* **63,** 553–602 (1964).
2. Shteinberg, Ya.G. *Strontium Glazes* (Ed. and trans. by T. J. Gray). Kaiser Strontium, Halifax, N.S., Canada. (1974)
3. Gray, T. J. *Bull. Amer. Ceram. Soc.* **58,** 768–770 (1979).
4. Marquis, J. E. and Eppler, R. A. *Bull. Amer. Ceram. Soc.* **53,** 443–445/9 (1974).
5. Baudran, A. and Ducare, R. *Interceram.* **23,** 64 (1974).
6. Bull, A. C. *Trans. Brit. Ceram. Soc.* **81,** 69–74 (1982).
7. Batchelor, R. W. *Trans. Brit. Ceram. Soc.* **73,** 297–301 (1974).
8. Beechan, C. R. and Williamson, W. O. *Trans. Brit. Ceram. Soc.* **62,** 939–947 (1963).
9. Berg, P. W. *Science of Ceramics* (Ed. G. H. Stewart) vol. 1, pp. 51–61 (1962).
10. Eppler, R. *Proc. 3rd International Cadmium Conference, Miami,* p. 31 (1981).
11. Brit. Pat. 1,522,868.
12. U.K. Pat. GB 1,403,470.
13. Knizek, J. *Trans. Brit. Ceram. Soc.* **59,** 339–387 (1960).
14. Vogel, W. *Structure and Crystallisation of Glasses,* p. 71 Pergamon (1971).
15. *Zircosil in Glazes,* Anzon Limited.
16. Olby, J. K. and Laidler, D. S. Quoted in Booth, F. T. and Peel, G. N. *Trans. Brit. Ceram. Soc.* **58,** 532–564 (1959).
17. Sehlke, K. H. L. and Täuber, A. *Trans. Brit. Ceram. Soc.* **68,** 53–56 (1969).
18. Swann, J. D. *Bull. Amer. Ceram. Soc.* **48,** 763–765 (1969).
19. Jacobs, C. W. F. *J. Amer. Ceram. Soc.* **37,** 216–220 (1954).
20. Biffi, G., Ortelli, G. and Vincenzini, P. *Ceramurgia International,* **1,** 31–35, (1975).
21. Biffi, G., Ortelli, G. and Vincenzini, P. *Ceramurgia,* **5,** 3–12 (1975).

22. Private communication, Cookson Ceramics & Antimony Ltd.
23. Brit. Pat. 1,436,859. *Glazing Composition*, Tioxide Group Ltd.
24. Norton, F. H. *J. Amer. Ceram. Soc.* **20,** 217–224 (1937).
25. Lane, R.O. *Ceram. Ind.* **37,** 62–68 (1941).
26. Zahn, W. A. *Ceram. Eng. Sen. Prog.* **2,** 943 (1981).
27. Brit. Pat. 1,436,859.
28. Hackler, C. L. *Bull. Amer. Ceram. Soc.* **59,** 647 (1980).
29. Skuthan, R. *Interceram,* **26,** 52 (1977).
30. Marquis, J. E. and Eppler, R. A. *Bull. Amer. Ceram. Soc.* **53,** 443–446 (1974).
31. Buldini, P. L. *Bull. Amer. Ceram. Soc.* **56,** 1012–1014 (1977).
33. Marquis, J. E. and Eppler, R. A. *Bull. Amer. Ceram. Soc.* **53,** 443–448 (1974).
32. Clough, A. R. J. and Simcock, H. *Trans. Brit. Ceram. Soc.* **75,** 36–39 (1976).
34. Harman, C. G. *J. Amer. Ceram. Soc.* **27,** 231–233 (1944).
35. Brit. Pat. 1,362,626. CSIRO.
36. U.S. Pat. 3,499,787; Brit. Pat. 1,107,943.
37. Maki T. and Tashiro, T. *J. Ceram. Assoc. Japan,* **74,** 89 (1966).
38. U.S. Pat. 3,561,984 and 3,565,644.
39. Brit. Pat. 1,239,328.
40. U.K. Pat. Application GB 2,080,790.
41. Duke, A., Megles, J. E., MacDdowell, J. F. and Bopp H. F. *J. Amer. Ceram. Soc.* **51,** 98 (1968).
42. Roberts, G. J., Salt, S., Roberts, W. and Franklin, C. E. L. *Trans. Brit. Ceram. Soc.* **55,** 553–602 (1956).
43. Jackson, W. *Ceramic News* **13,** 15 (1964).
44. M. Vukovich, Jnr., *J. Can. Ceram. Soc.* **31,** 100 (1962).
45. *Zircosil in Glazes,* Anzon Ltd., p. 41.
46. Turner, W. E. S. *J. Soc. Glass Tech.* **40,** 276–299 (1956).
47. Leach, B. *A Potters Book,* p. 162 Faber & Faber (1953).
48. Forrest, J. S. *J. Sci. Instrum.* **24,** No. 8, p.211 (1947).
49. Forrest, J. S., *J. Instn. Elect. Engrs.* **89** (2), 50 (1942).
50. Smith, E. J. D. *Trans. Brit. Ceram. Soc.* **58,** 277–300 (1958–9).
51. Binns, D. B. *Trans. and J. Brit. Ceram. Soc.* **70,** 253–263 (1971).
52. Binns, D. B. *Trans. and J. Brit. Ceram. Soc.* **73,** 7–17 (1974).
53. Binns, D. B. *Trans. and J. Brit. Ceram. Soc.* **70,** 253 (1971).
54. Taylor, R. H., Allinson, D. L. and Barry T. I. *J. Mater. Science,* **13,** 876–884 (1978).
55. Binns, D. B. *Trans and J. Brit. Ceram. Soc.* **73,** 7–17 (1974).
56. Dimarcello, F. V., Treptow, A. W. and Baker, L. A. *Bull. Amer. Ceram. Soc.* **47,** 511–516 (1968).
57. ASTM, Specification for ceramic glazed facing tile and brick.
58. Verba, R. J. *Bull Amer. Ceram. Soc.,* **37,** 364–365 (1958).
59. Brit. Pat. 739,971 (2/11/55).
60. Tauber, E., Crook, D. N. & Murray, M. J. *Constructional Review,* **43,** 58–65 (1970).
61. Tauber, E., Murray, M. J. and Crook, D. N. *J. Aust. Ceram. Soc.* **7,** 18–24 (1971).
62. Brit. Pat. 1,176,755.
63. ILZRO, *User Manual for Flame Spraying of Lead Glass Coatings.*
64. Rado, P. *Interceram.* **12,** 140 (1964).

Chapter 8
Glaze Application

At the point where glaze and ceramic substrate come together, much value has been added to each component. It is, therefore, essential for the application of the glaze to be straight forward, avoiding errors which could lead to waste. The method must be reproducible, economical and flexible.

Each type of a glaze has its optimum thickness of application. In the unfired state this will vary from about 0.15 mm to 2.0 mm. As a general rule the glaze will be applied as thinly as is necessary only to give a total uniform smooth coverage. If undercoated, respraying can add thickness where required, but reduction in glaze thickness is not easy. There is a reluctance for glaze while firing to flow and heal across thin or starved patches or to level out irregularities such as ripples or runs in the coating. The glost fire will not balance colour variations due to differences in glaze thickness.

How the glaze is to be applied is a most important question to answer, and there are a number of different methods of application which can be chosen. The selected method will only be approved after considering several important criteria.

1. Type of ware.
2. Shape and size of ware.
3. Throughput.
4. Energy sources.
5. Labour.
6. Space.

Lead ore, alone or mixed with some flint or clay, was once applied as a glaze. In the old method the glazing, material was powdered galena, which was dusted through an open weave cloth bag onto the slip-coated body or onto the newly thrown clay. In the kiln interaction between lead ore and clay, the outer skin of the ware was fused into a vitreous coating of variable quality and poor colour. Later the same glazing compositions were applied by dipping biscuit ware, without much improvement in quality. Nevertheless, this early technology provided the base from which consistent glazes developed.

8.1 DIPPING

The commonest method, dipping, is a simple, efficient and quick technique, requiring no special plant, although success with the method depends upon experience.

Dipping is the operation where the ceramic is immersed in the glaze slip, withdrawn, allowed to drain and dry. There is an optimum thickness for any glaze layer which lies within narrow limits and the several sections of any application method are closely monitored to ensure that the glaze is deposited at the optimum thickness. Once this is achieved, the quality of the coating must be considered for its effect on the surface which develops in the glost fire. The amount of glaze taken up depends upon the state of the ware, the rheological properties of the slip and the time of immersion.

The equipment needed is not extensive. In its simplest form a receiving tank or "tub" large enough to contain a depth of slip sufficient to permit the dipper freely to immerse the ware is required. Refinements include an automatic recirculating and cleansing system which continually takes slurry from the tub and discharges it back through a sieve and an electromagnet for re-use.

How dipping is executed depends upon the shape and size of the ware. Hollow-ware and small flatware lightly held by the foot and rim is completely immersed in the slip for a few seconds, with a gentle to and fro agitation. Flatware having a diameter greater than a hand span is gripped by the fingers and a wire hook on the dipper's thumb. The ware is taken from the slip, shaken to remove excess glaze and set down to drain. Immediately any bare or thinly glazed patches are touched-up with a finger wet with glaze.

The surface texture of the fired glaze is the result of many interrelated factors. These are:

1. Fluid density.
2. Viscosity of glaze slip.
3. Thixotropy of the slip.
4. Fineness of grind of the glaze.
5. Operator expertise.
6. Porosity of ware.
7. Thickness of the ware.
8. Temperature of ware.
9. Time of immersion.
10. Presence of hardeners, starches, etc., in the slip.

Compensating action for some of these factors can be taken. Adding to the immersion time does not necessarily give a heavier take-up of glaze. Glaze thickness will depend upon other factors such as biscuit porosity, fluid density and thixotropy of the slip. Nevertheless, variations in ware thickness can cause variations in glaze thickness.

Glazes for highly porous ware will have a lower slip density than glaze for harder fired biscuit and they are therefore more easily applied. As an alternative the porous biscuit can first be dipped in water before immersion in a high-density slip. Vitreous ware requires a high-density glaze slip with a high thixotropy; consequently, there is a tendency for pinholes to appear in the coating.

Slip density of glaze appropriate for dipping ranges from about 1.4 kg l^{-1} for clay ware to 1.7 kg l^{-1} for biscuit ware. Vitreous bodies require higher slip densities in the region of 1.7–1.9 kg l^{-1}.

Double dipping is a glazing method where the inside and the outside of the piece of hollow-ware are coated simultaneously, leaving the base of the ware unglazed. The ware is dipped into the glaze, rim first, and with a sharp jerk of the hand upwards the inside is splashed by the agitated glaze.

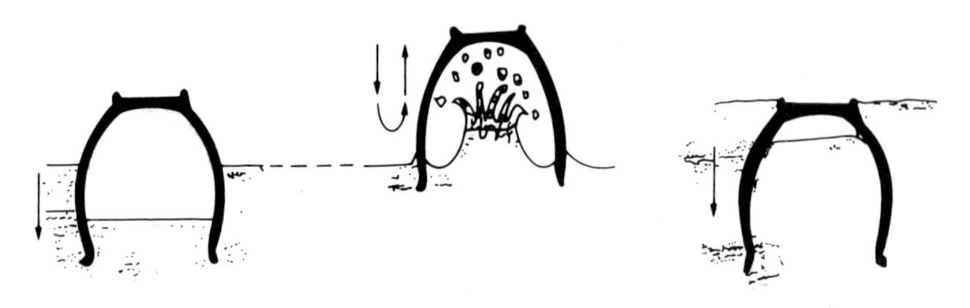

Fig. 8.1.

With the inner surfaces coated, the ware is pushed downwards into the slip until it is fully immersed and the outside is also completely glazed.

The principle has been mechanised. A reservoir has in its base a vertical pipe by which glaze can be pumped with enough force to raise the jet by a few centimetres. The flow of glaze is controlled through a foot-operated pedal. In the upper part of the reservoir is a tray which positions the hollow-ware at a fixed distance from the nozzle. This tray may or may not be below the surface of the glaze slip. When it is below the surface, the ware rests on the tray, coating the wall to a fixed distance below the rim edge. With the cup inverted the pedal actuates a spray of glaze within the ware and the inside receives a uniform coating. The outside of the ware is glazed in a separate operation.

8.1.1 Automatic Dipping

This system is suited to the glazing of porous biscuit. With the techniques of hand dipping understood, the difficulties of reproducing the process mechanically are obvious. These difficulties centre on the method of gripping the ware—usually flatware—and holding it uniformly in the glaze.

The quality of the application is determined by:

1. Porosity of the ware.
2. Fluid density of the glaze slip.

3. Thixotropy of the glaze slip.
4. Time of immersion.
5. Temperature of the ware.

In the simplest scheme, flatware drops smoothly along an inclined support until fully immersed in the glaze. In a continuous movement it is drawn out on the moving wire belt. In a more complex system, suction heads mounted on arms hold the ware and introduce it to the glaze reservoir in two separate stages. At the first station the back of the ware is glazed and the foot cleaned. The ware is then transferred to another set of suction heads, inverted and passed through a second glaze reservoir. Drying and placing follow in a continuous sequence.

8.2 SPRAYING

With suitable equipment, glaze slip can be broken down into a cloud of fine particles which are transferred to the ceramic body by either pneumatic, mechanical or electrical forces. Each process has its own range of controls to give the best type of coating in the most economical way. The method requires a "gun", a container or feed mechanism, an impelling agency and a properly designed hood or booth maintained under negative pressure. Spraying requires only small quantities of glaze. Thick or thin coats can be applied to substrates which vary in their degree of vitrification or porosity. Large or intricately shaped ware can be uniformly coated.

Loss of glaze by overspray and by rebound is a factor for consideration, because this could be as high as 50% of the feed material. This material can be collected, reprocessed and recycled.

8.2.1 Air Spraying[1]

The equipment needed is a spray gun (of which there are several types), a compressor and a spray booth.[2]

Slip flows through a control orifice or nozzle in the gun where it meets a stream of pressurised air. This air stream disrupts the flow of liquid into a myriad of fine particles. A third air stream might be used to direct the glaze globules into a cone-shaped cloud. The flow of slip and air is controlled by a trigger and the volume of slip is regulated by the size of the nozzle. The aperture is generally of a larger diameter for viscous glaze slips. As the aperture wears, the flow rate changes, consequently for abrasive materials such as glaze, wear-resistant nozzles are fitted.

There are three main types of feed, whose usefulness depends upon the application. These are gravity, suction and pressure feed.

1. *Gravity.* This system is best used for small trials rather than production. From a small cup attached to the gun or from a separate container linked by a tube the slip flows to the nozzle. The rate of flow depends upon the viscosity of the slip.

2. *Suction.* The nozzle is supplied from a cup attached to the gun from which the slip is sucked. Its applicability is limited because of the small size of the container and because of difficulties inherent in viscous and dense slips.

3. *Pressure.* With the most viscous slips, a pressurised feed is the preferred option. Slip is positively fed to the nozzle usually from a separate container set below the level of the gun. Some systems are fitted with insert containers to permit a rapid changeover to other glaze compositions. Agitators are fitted to the store tank but since most glazes are stable they require only an occasional stir.

8.2.2 Parameters for Spraying

The quality of the fired glaze as it leaves the glost kiln can be related to the quality of the sprayed surface on the ware as it enters the kiln. Factors which affect the uniformity of the sprayed surface partly originate in the glaze slip and partly in the mechanical adjustments of the plant, yet all are interrelated. The factors are:

1. Slip density.
2. Viscosity of the slip.
3. Thixotropy of the slip.
4. Fineness of grind of the glaze.
5. Air pressure.
6. Air flow.
7. Feed rate of glaze.
8. Nozzle aperture.
9. Operator expertise.
10. Degree of body vitrification.
11. Temperature of the article (when preheated).
12. Position and number of guns.

Two types of glaze finish can be produced according to whether the technique of "dry" spraying or "wet" spraying is used. Under different conditions the atomised spray can arrive at the ware either having retained most of the water within the droplets or having lost some of the moisture in transit. When the droplets coalesce on the ware the deposit is consequently "wet" or "dry". A "dry" spray can be produced by increasing the amount of atomising air or reducing the flow of slip to the nozzle. A high-density or high-viscosity slip usually produces a "dry" surface. Such a technique produces a dry powdery deposit of glaze whose inherent roughness and porosity means that higher firing temperatures or longer peak soaking times are needed for full glaze development. The usefulness of this technique is where there is a need either to build thick coatings without sagging, to coat awkwardly contoured shapes without cracking or to cover underglaze decoration without the need for "hardening-on". A similar effect can be achieved by preheating the ware prior to spraying.

Glazes applied by a "wet" spray have a smooth surface texture before firing, consequently a high gloss develops at lower kiln temperatures. "Wet" spray on a porous body can offset the adverse effect of the rapid take-up of moisture from the spray. For normal spraying the glaze is applied until the surface acquires a slight glossiness or wetness. Beyond this point the glaze will either sag or it will ripple under the impact of the atomised spray. Slips with low fluid densities are more inclined to "orange peel", but this can be countered by reducing the feed rate. Changes in the nozzle size and atomising air will alter the efficiency of fine droplet production. The size of the droplets can be increased to the point where small splashes will "stipple" the ware.

8.2.3 Automatic Spraying[3]

For high volume tableware and tile production, glaze spraying is automated. The articles are continuously fed onto a two-wire belt and through a preheat tunnel before the glaze is applied from a battery of angled spray-guns such that the underside is first coated. After inverting the ware the top face of the ware is similarly coated with glaze prior to passage through a second dryer. The ware can then be placed immediately for glost firing. Either biscuit or clay ware can be coated. Glaze is circulated through sieves and magnets before reaching the traversing guns. As little as 2% of the glaze will be retained on the ware, leaving a large volume of slurry to be recycled.

Glaze reclaim is essential for economic management. There are three sources of the recovered glaze.

1. Slip caught by rear baffles or "curtain" within the booth.
2. Over-spray transported by air-currents to adjacent dust control equipment.
3. Washings from cleaning operations.

As a high volume repetitive production operation, spraying is ideal for robotised automation. The robot and its computer systems can be integrated with a standard spray system. Robot spraying machines are "taught" by a skilled operator who guides the robot through the sequence of spraying movements and trigger operations. This sequence is recorded as the program. Several different programmes can be recorded for different shapes of ware. The spray gun is attached to the arm of a robot arm which is actuated by hydraulics and capable of movement in three directions. Economy in glaze usage up to 40% is obtained.

8.2.4 Mechanical Spray[4, 5]

An alternative to the use of air to provide atomisation is where the spray is produced mechanically with a rotating atomiser. The mist of glaze deposits only a thin but uniform coating. Glaze slip is passed through a hollow spindle onto a set of rotating discs, slightly spaced apart. Centrifugal force throws off the glaze into a fan of droplets. The ware passes through this spray at a speed depending upon the desired coating thickness. Different applications can be made by altering the diameter and speed of rotation of the discs and so modifying the characteristics of the fans. A curved hood encloses the discs to collect overspray. By balancing the distance the hood is away from the discs, the overspray on the hood never dries and faults, due to dry glaze particles falling onto the ware, are eliminated.

A mottled texture to the glaze can be applied by changing the spindle conformation, replacing the discs with cups. The cups rotate vertically throwing glaze over the castellated rim to produce large droplets whose size can be varied as required.

Slip densities in the region of 1.7–1.8 kg l^{-1} are used for this type of spraying.

8.2.5 Electrostatic Spray[6, 7]

Application by electrostatic spray produces a high surface quality and the loss of glaze from overspray is reduced. Overspray represents a high proportion of glaze in circulation in air-based spraying. Electrostatic spraying requires a conductive body to act as "earth" in the circuit, consequently green ware more easily receives a uniform coating than does a biscuit body. Additives modify the electrical properties of the slip and hardners are necessary.

Glaze reaches the applicator "gun" from a pressurised stock container through a regulator. Droplets are produced either by air atomisation or by centrifugal force from a sharp-edge rotating surface. In the mechanical atomiser a disc spins at high speed and glaze, fed to its centre, is thrown from the edge. The drops acquire a high negative potential and as the like-charged particles repel each other they are dispersed into a fine mist. At the same time they are driven forward to the earthed ware following the lines of force. Glaze can reach the underside of the ware, and full edge coverage is achieved as a homogenous cloud develops around the ware.

There is a back spray effect from particles rebounding from the ware. On impact, particles lose their charge and on rebounding they become recharged sufficiently to travel back to the spray equipment. By placing a ring of ionisers around the gun this can be nullified. Additionally jets of air around the spray head can redirect glaze droplets back to the ware.

An electric eye and programme controller automatically trigger the guns when ware arrives at the spray station. By having a number of guns operating at the same time each gun applies only a proportion of the coat, therefore overloading the guns cannot occur.

The environment in the spray booth plays an important role in the success of this method. Temperature and relative humidity are maintained within narrow limits because of their effect on the electric field and the condition of the glaze spray. When the humidity is low, then the glaze slip loses water and the glaze spray arrives at the ware "dry". Low air speeds only are generated and additional control of hygiene must be exercised.

All types of glaze can be sprayed, provided the rheology of the slip is adjusted. For leadless glaze with a slip density of 1.5 kg l^{-1} the viscosity of the slip will lie between 20 cp and 500 cp.

8.3 SEDIMENTATION AND TAPES

Some microelectric substrates use vitreous coatings as an integral part of the composite. These coatings can be classed as glazes. The methods of producing precisely defined coatings are specific to the technology.

8.3.1 Sedimentation

Glaze films used for the surface passivation of transistors and diodes can be applied by sedimentation. When this process is accentuated by centrifuging, thin, uniform coatings free from pinholes are deposited. Both low firing and high firing glazes have been used.

A silicon wafer or alumina substrate is placed on the flat base of a cylindrical centrifuging cup which contains a suspension of finely ground frit (Fig. 8.2a). The particle size of the frit is smaller in diameter than the required coating thickness which is approximately 0.5–1.5 μm.

Conventional glaze grinding produces only a minor proportion of small diameters and correctly sized particles can be separated from the glaze slip by prior sedimentation or centrifuging.[8] This "cut" is then redispersed in the suspending liquid over the substrate. If this suspension is centrifuged at high speeds for a fixed time, a uniform layer of fine particle is deposited on the substrate. (Fig. 8.2b). The supernatant liquor is decanted (Fig. 8.2c).

The substrate is removed and fired at a temperature whose value depends upon the frit composition, its particle size and the layer's thickness. Using a two-coat process with an intermediate firing gives pinhole free coatings.

The viscosity of the suspending medium should be low to avoid disturbing the sedimented coating when decanting. It should also be volatile so that through drying of the deposited layer is not inhibited. When making the choice of a suitable medium, the particle size of the frit and electrical properties of the liquid are considered together.[9]

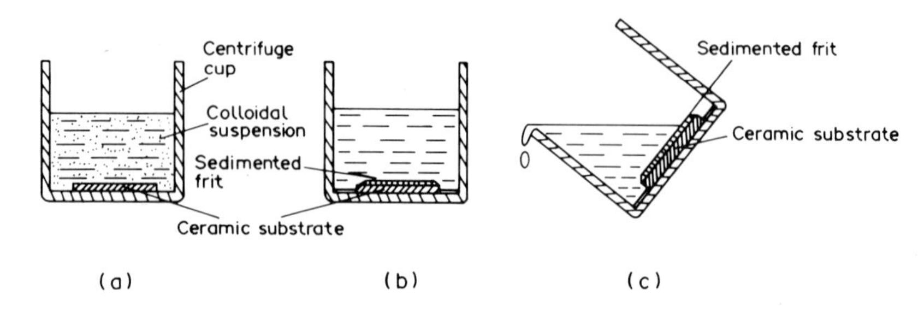

Fig. 8.2. Sedimentation of colloidal frit

8.3.2 Transfer Tapes

Tapes are used for coating a substrate with a layer of glaze with precise thickness and density, and as a coating method, can replace sedimentation. A glaze frit powder layer is attached to a plastic carrier film. This backing material is thin, flexible and uniform. After deposition on the tape the frit is covered with a layer of a pressure-sensitive adhesive. This sandwich is protected by a release paper.

Using this technique, thin coatings of glaze frit, about 5 μm and less, can be produced. Any surface shape can be glazed by pressing the tape with uniform pressure into place even on curved surfaces.

The composition of the frit is selected for type (vitreous or devitrifying) and melting point. Firing cycles cover temperatures from 460°C to 1300°C. If the firing is in air, no change is necessary in the schedule, but if neutral or reducing atmospheres have to be used because of the nature of the substrate, then a preglazing fire at 400°C for 15 min is necessary to volatilise all the organic adhesive.

8.4 CURTAIN COATING

Tiles require only one face to be glazed, and coating by "waterfall" or "curtain" is the commonest method. A continuous feed of tiles carried on parallel tracks of stranded wire or rubber belt passes under a curtain of fluid glazed slip. The slip continuously circulates in the machine, flowing from an overlying reservoir in a controlled curtain into a lower sump, from which it is pumped through a filter into the upper reservoir.

As the leading edge of the tile enters the curtain the glaze flows across the horizontal surface in a uniform manner and only the face of the tile is glazed. The speed of traverse through the waterfall, the density of the glaze slip and the porosity of the body control the thickness of the coating retained by the tile. To limit the occurrence of bare spots or thinly glazed areas, the coating waterfall unit is usually duplicated.

The coating varies in thickness from the leading to the trailing edge. With thin deposits this unevenness is too obvious and to counteract the thin front edge, the tile is passed twice under a curtain of glaze, turning the tile through 180° between each coating its equalise the weight. Very porous biscuit is dampened by a spray of clean water before reaching the glaze curtain to reduce glaze take-up. Vitreous biscuit needs to be preheated before coating.

Sundry treatment facilities are integral with the coating unit. The feed automatically distributes tiles top face uppermost onto the conveyor. Dust is removed from their faces as the tiles pass under rotating brushes. At this point in the system a silk screen decorating machine can be stationed. Different glazes can be applied by marbling rollers or by random sprinkling of contrasting colours. A preheating tunnel might be installed if the tile body has only limited porosity. Following the curtain applicator, the tiles can be fettled. Extra stations for drying and second stage decoration can be fitted.

8.5 TRAILING

Though not strictly a glazing method, trailing is used to apply glaze as part of a decorative process. The glaze powder is dispersed in an aqueous-based medium, thickened with a carboxymethyl cellulose or similar viscosity agent. A preliminary, dipped coating of glaze is applied and before it is completely dry, decorative lines are piped on with a bulb trailer.

8.6 PASTE

Coloured glaze having an increased clay content is made into a paste[10] with a small amount of water to which a deflocculant has been added. One or more of these varicoloured slips is applied firstly to a flat steel plate as a pattern. The pattern is transferred to a plaster mould into which plastic ceramic body is pressed or cast, so producing a decorative glazed ceramic surface. Alternatively, the paste is applied directly to the plaster mould. A single fire matures both glaze and body.

8.7 PRESSING

As a modification to an old technique, floor tiles receive a coating of glaze in a press.[11] Spray-dried glaze in a form similar to that of the body "dust" is fed first to the press mould before the fill is completed with body. An alternative procedure[12] reverses the order of filling. The tile is pressed and the compaction of body and glaze occurs together. The ceramic/glaze unit is once-fired.

8.8 POURING

When there is an insufficient volume of glaze for dipping, then pouring is often the only possible method of glazing.

Pouring is a viable method of glazing large, or narrow-necked, ware and it is more in use by the studio potter than by the industrialist. For large ware the interior is coated and then, with the ware inverted (and preferably on a turntable), glaze slip is poured over its outer surface. A variety of effects allied to the draining characteristics of the glaze slip is possible and many such textures are encouraged by the craft potter.

8.9 PAINTING AND BRUSHING

Now an uncommon technique, but it can be used by the craft potter for special effects and for applying glazes to places inaccessible to the spray gun. Only a small amount of glaze is needed.

The glaze slip is modified by the addition of a binder to increase adhesion to the substrate and by the addition of an emollient such as glycerine to give a smoothness to the coating. Several thin coats are applied and dried between each application to deposit the correct weight of glaze on the object rather than attempt one thick coat. "Mop" brushes (with soft hair) are fully loaded with glaze and with sweeping movements made over the ceramic surface (to prevent plucking or dragging) overlapping swathes of glaze are deposited. Painting onto fired glaze and refiring is a method of reclaiming substandard ware.

Tiles are decorated with glaze by brushing. After passing under a waterfall applicator, the wet surface is touched lightly with a brush coated in glaze so that the fired tile has a streaked finish. Machines are available for traversing the brush across the tile in a predetermined pattern and when several brushes are used the interwoven patterns are complex and non-repetitive.

8.10 ELECTROPHORESIS

Finely-divided vitreous compositions can be applied electrophoretically.

This process is the electrical deposition of glaze material from a water suspension which is an integral part of an electrical circuit. The material in suspension acquires a positive charge and is deposited on the negatively charged work piece. The container forms the anode. The particles move at differing speeds depending upon their net electric charge, size and shape. Frit and colouring agent move at different rates, which gives shade variations; consequently, self-coloured frits are preferred.

The principal advantage of the method is in the uniformity of the coating thickness, even on complicated shapes. The deposit on the ware is densely packed.

8.11 FLAME SPRAYING

Glassy coatings can be applied in the molten state to heat-sensitive or massive substrates. Such coatings obviate the need to heat up the bodies to develop vitrification in the glaze. The technology is well established but not widely applied to glazing. Though surfaces of a satisfactory quality can be obtained, these do not match those from dipping or water-based spraying.

There are three stages of equal importance to success:

1. Heating the glaze in a flame (or plasma) until molten.
2. Projecting the heated particles at high velocity.
3. Depositing the hot particles as an adherent coat.

Powdered glaze is fed to the flame in the hot zone of a spray gun. Flame guns use acetylene or hydrogen and oxygen as their energy source. Particles have velocities of 30–45 m s^{-1}. Some particles are not completely molten even though temperatures are high.

The quality of the coating depends[13] upon the viscosity, specific heat and thermal conductivity of the molten powder and on the thermal conductivity of the substrate. This latter function determines the flow in the glaze before solidification. A principal difficulty is in supplying the powder at a uniform rate. Particle size distribution is important if all grains are to melt. The viscosity[14] of the glaze at the melting point is vital to the development of a smooth surface since it will determine how the molten grains will adhere and coalesce. The distance between substrate and gun nozzle is critical. For short distances, complete melting of each granule might not have occurred and, for longer distances, the particle might have partially solidified.

8.12 SOL-GEL GLAZING

The sol-gel method of glass-forming[15] does not involve melting crystalline raw materials. Many glasses can be prepared by this novel method with compositions similar to those made by melting the constituent oxides at high temperatures and then cooling. "Sols" are suspensions of very fine solid particles such as colloids in solution. Increase in the viscosity of a sol, usually by partial loss of the liquid phase, results in the formation of a rigid "gel".

Mixtures of liquids are reacted, the product of which after subsequent dehydration gives a solid. Residual organic and hydroxyl groups are then removed and the solid is vitrified by heating to form a glass at temperatures which are below the glass transformation temperature, T_g.

There are advantages over the conventional glass melting route. Homogeneity and purity are better. The preparation temperature is lower even for new compositions previously outside the normal range of glass formation. Films can be formed *in situ*. The main disadvantage is cost, as well as the drawbacks of the large shrinkages which occur, the presence of fine bubbles and the long processing times. There is an obvious application in electroceramics for these high-grade glazes.

A thin glazing coat is formed directly on a substrate[16] from a solution or emulsion of the precursors of the glaze-forming oxides. The precursors can be salts of organic acids dissolved in alcohol. This liquid is applied to the ceramic substrate. It is dried and heated. As the temperature rises, the solvent distils and the organic acid radicals decompose, leading to the formation of a thin glassy layer. The layer can be made conductive by including precious metal compounds in the reactive liquids.[17]

The technique can be adapted to produce protective glaze coatings on porous silicon carbide and silicon nitride ceramics.[18] The reacting liquids containing germanium and silicon infiltrate the ceramic under reduced pressure. Completion of the glass forming action takes place in the temperature range 700–1000°C.

8.13 SILK SCREEN

For substrates used in electronics which require precisely positioned areas of glaze, the silk screen process is used.

Finely powdered dry glaze or frit is dispersed in a medium to form a smooth paste. Using a sharply-edged rubber squeegee this paste is pressed through the open areas of a fine mesh stretched on a frame. Once this mesh was of silk, but now it is either metal or polyester. The mesh is coated with a photosensitive emulsion and then dried. A transparent "positive" opaque black ink drawing of the design is pressed close to the screen prior to exposing the emulsion to strong light. Exposure to light desolubilises the emulsion. The mesh is washed in warm water to leave the pattern *in* the filaments of the weave.

The substrate is positioned below the screen and as the two contact, the paste adheres in a precise pattern relating to the open area of the design in the screen. Control over the thickness of coating is precise through adjustment of the paste density, diameter of the mesh filament, thickness of the plastic blocking out agent, the distance of the screen from the substrate and pressure on the squeegee. As firing proceeds, the organic components of the vehicle burn away, leaving no residue, and the glaze will mature at its designed temperature.

The same technique is used in several decorative processes for tiles and tableware of which glaze is an integral part. Instead of having an overall coating, selected areas of the biscuit ware can be coated. Exact positioning is paramount and special jigs register the substrate accurately each time under the screen.

The medium in which the glaze powder is dispersed can be based on oil or wax mixtures. With wax-based printing mediums the "inks" are kept liquid on the screen either directly by heating a metal screen (by passing through its filaments a small current) or indirectly by heating a polyester filament screen (by infrared radiators positioned above). The inks solidify instantaneously on contact with the cold surface of the ware. The ware can be immediately further treated by the application of more glaze by other techniques. Oil-based paste depositions of glaze need drying if they are to receive further treatment without rejecting subsequent coatings of glaze in aqueous suspension.

Compositions of glaze powder and organic medium can be made with a high degree of water repellency.[19-22] These are used as components in the "resist" printing process for multi-colour decoration. When the printed ware is dipped in a water suspension of glaze of the ink deposit repels the slip to the borders of the printed area. In the glost fire there forms a continuous coating of the contiguous deposits.

For special effects, standard glaze compositions can be modified to increase or decrease the degree of reactivity with other glazes at high temperatures. As the firing proceeds, the glaze powder will either mature normally or it will react with the adjacent glaze to form variable effects.

8.14 "SELF GLAZING"

Cordierite bodies are difficult to match with a glaze having an adequately low thermal expansion, but some bodies of this type develop, during firing, a very thin glassy layer equivalent to a glaze.[23, 24] The thickness of this film is about 0.025–0.05 mm and it is coherent with the body through a well-defined buffer layer of cordierite crystals.

By adjusting the composition of the body to produce a cordierite structure and then adding a proportion of fluxing minerals such as felspar, nepheline syenite and zinc oxide, tableware articles can be formed which self-glaze at 1350°C. Matching slip-glazes have been made.[25] This self-glazing property can be enhanced by the addition of transition metal oxides.[25, 26]

8.15 FLUIDISED BED[27-30]

A process capable of development and automation is the fluidised bed method. By this technique glazing can be achieved independently of the properties of the substrate.

In a fluidised bed air is passed through a bed of closely graded powder. A stream of air moves from a plenum chamber via an air distributing plate at a velocity high enough to suspend the grains. Under these conditions the bed expands and behaves as a fluid in a semi-stable condition. Glaze powder which is substantially glassy is used as the bed material. Its particle size is less than 110 μm and ideally the grains are as near spherical as possible. To this glaze powder is added a particulate resin. With the ware preheated to a temperature above the melting point of the resin (about 200°C, but perhaps up to 350°C), it is dipped into the fluidised bed.

The immersion resembles conventional dipping and the ware is held in the bed long enough to acquire an adequate coating. As slip density is a factor in the dipped thickness of a glaze coating, so is the ratio of glaze to resin a factor in this technique. The optimum addition lies within the range 1–5% resin (based on weight of glass). Other factors are the preheat temperature and glaze particle size.

In firing, the resin volatilises before the glaze powder softens. The quality of the glaze coating is good.

REFERENCES

1. Whitmore, M. *Trans. Brit. Ceram. Soc.*, **73,** 125–129 (1974).
2. Bloor, W. A. and Eardley, R. E. *Trans. Brit. Ceram. Soc.*, **77,** 65–69 (1978).
3. Whitmore, M. *Trans. Brit. Ceram. Soc.*, **73,** 125–129 (1974).
4. Cimes SNC Automazioni Industriale, via A Ascari 21/23, 41053 Maranello, Modena.
5. Omis.
6. Hebberlein, K. *Ber. Deutsch. Keram. Ges.*, **53,** 51–55 (1976).
7. Lambert, M. *L'Indust. Cer.* 376–380 (1974): *Interceram,* **23,** 107 (1974).
8. Johnson, R. *Trans. Brit. Ceram. Soc.*, **55,** 267–285 (1955–6).
9. Pliskin, W. A. and Conrad, E. E. *Electrochemical Technology,* **2,** 196–200 (1964).
10. Hurley, D. C. U.S. Pat. 4,294,635.
11. Faust, W. H. *Interceram.* **23,** 32–34 (1974).
12. Emiliani, T. and Biffi, G. *Ceramurgia,* **6,** 125–128 (1976).
13. Ault, N. N. *J. Amer. Ceram. Soc.*, **40,** 6–74 (1957).
14. *High Temperature Inorganic Coatings,* Ed. John Huminik. Reinhold, New York (1963).
15. Geffcken, W. and Berger, E. Deutsch Reichspatent 736,411 (1939).
16. Brit. Pat. 1,166,991, Plessey Co. Ltd.
17. Brit. Pat. 1,166,992, Plessey Co. Ltd.
18. Schlichting, J. Z. and Neumann, S. *J. Non-cryst. Solids,* **48,** 185–194 (1982).
19. Brit. Pat. 1,023,764.
20. Brit. Pat. 1,348,651.
21. Brit. Pat. 1,409,720.
22. Brit. Pat. 1,415,833.
23. Thurnaver, H. *J. Amer. Ceram. Soc.*, **20,** 368–372 (1937).
24. Brit. Pat. 1,362,626.
25. Theiss, L. E. *J. Amer. Ceram. Soc.*, **26,** 99–102 (1943).
26. Brit. Pat. 681,446.
27. Eastwood, M. C. Brit. Pat. 1,370,288.
28. Eastwood, M. C. Brit. Pat. 1,492,532.
29. Eastwood, M. C. Brit. Pat. 1,569,709.
30. Roberts, W. *Trans. Brit. Ceram. Soc.*, **73,** 47–50 (1974).

Chapter 9
Chemical Resistance

With examples to be seen of glazed pottery dating from several thousand years ago, it is understandable that the permanence of a glossy glaze seems to be unchallenged. Modern glazed tableware is immersed daily in hot alkaline water without its glossy surface disappearing, but glazes of all types are affected to some extent by contact with water, acid or alkali.[1] Sodium silicate is a glass which can be said to be "soluble" in water, yet water has little effect on vitreous silica. Water is an important factor in the weathering of glazed brickwork, tiles and electrical porcelain. Silica-based glazes are attacked by phosphoric acid and vigorously by hydrofluoric acid. Lead bisilicate is attacked by hydrochloric acid.

The resistance to attack by liquids is a property of glaze which is crucial, and many glazes with other desirable properties are unacceptable because they are susceptible to corrosion by mild alkaline solutions or by food acids. There can be wide differences in properties between glazes for tableware and sanitaryware and glazes for artistic effect on decorative ware. If it is not to lose an essential feature of its definition, a glaze must retain its imperviousness, though it might lose all of its gloss during its working life. Liquids stored in glazed vessels become contaminated with elements extracted from the glaze. In one notable case whisky was spoilt after storage in pottery due to the migration of about 2 ppm of aluminium, calcium and magnesium ions from the glaze. Fruit juice has extracted from glaze, ions of lead, cadmium and copper to the former's detriment.

Reaction at the glaze surface[2] can involve ion exchange, dissolution or absorption. The result of these reactions might be seen as a dimming of the gloss, thin-layer interference colours or even surface pitting and degradation. Attack on the glaze might release cations which, when they are part of the glaze structure, are safe, but which are undesirable in associated foods. Preferential solution of some part of the glaze might lower the quality of the surface and give rise to silver marking or tea-staining on tableware. Semiconducting glazes on electrical porcelain can fail in service due to changes in conductivity.

9.1 MODE OF ATTACK

Glass scientists have made much progress in understanding the interactions between silicate surfaces and liquids and those who are interested in glaze durability make great use of the results of the research.[3a, 3b, 3c]

Attack by liquids on glass is a complex process. It is not one of dissolution, but rather is it a penetration and interaction with the vitreous structure by some ions leading to the eventual decomposition of the glass. The reaction is rapid. When an alkali–silicate glass is in contact with deionised water, the water instantaneously becomes a solution of alkali and silica.[4, 5]

Because the majority of glazes contain silica it is convenient to consider reactions involving compositions based on a silica network. Attack begins with an interaction between the lattice ions and hydrated protons in the contacting solution. Small quantities of associated monovalent and divalent cations are consequently leached from the structure.

9.1.1 Attack by Water

The corrosion by an aqueous solution of a glaze surface containing silicon and alkali metal, once initiated, could be expected to follow a cycle similar to that proposed for glass.[5] Firstly there is the replacement of an alkali ion by hydrogen from water:

(i) $\equiv Si{-}OR + H.OH \rightarrow \equiv Si{-}OH + R^{+}OH^{-}$.

Secondly, the hydroxyl ion interacts with the siloxane bond (Si–O–Si) in the vitreous network:

(ii) $\equiv Si{-}O{-}Si\equiv + OH^{-} \rightarrow \equiv Si{-}OH + \equiv Si{-}O^{-}$

Thirdly, the open oxygen so formed interacts with another water molecule to produce a hydroxyl ion which becomes available for another reaction (ii) to repeat the cycle:

(iii) $\equiv Si{-}O^{-} + H.OH \rightarrow \equiv Si{-}OH + OH^{-}$

As H^{+} ions replace R^{+} ions there is produced a surface film resembling silica gel having different properties from the original mass glass. This film swells, acts as a barrier to further reaction and decreases diffusion rates of ions into and out of the surface, thereby inhibiting further attack. If, at an intermediate stage in the attack, this layer dries out, the resulting thin film of silica produces characteristic "interference" patterns. When developing chemically resistant glazes it is encouraging to see these irridescent fringes after a durability test because glazes exhibiting diffraction colours are usually found to be more resistant than glazes which show only a reduction in gloss.

9.1.2 Attack by Alkali

The appearance of alkali ions R^{+} in the attacking liquid adds to the complexity of the attack process. Attack is no longer by "water", but comes from a liquid containing a number of elements whose proportions are related to their various extraction rates.[8] A glass containing sodium oxide would under these conditions release, as well as silica, sodium ions for further reaction. In the case of a silicate glass this could be represented as

(iv) $\equiv Si{-}O{-}Si\equiv + Na^+OH^- \rightarrow \equiv Si{-}OH + \equiv Si{-}O^-Na^+$

The presence of alkali ions in the solution will increase the pH of the solution. The rate of silica extraction under these conditions increases with the rise in the pH above 9[9]. Whilst the increase in pH might decrease the exchange of alkali ions from a soda-based glass, the additional alkalinity favours the dissolution of silica. Further alkali is then released through a breakdown in the silica network. Attack by alkaline solutions continues because no protective film forms on the surface of the glass.

The level of alkali extracted can be used as a measure of the inertness or reactivity of glaze to chemical attack.[10, 11] This extraction rate varies as the square root of time and it is concluded that the movement of the alkali ions is one of diffusion. For other glaze compositions, any ion released in the interaction of leachant liquid and glaze can be used to monitor durability, and many ions have been defined as acceptable monitors in health control regulations. An example of this is the international interest in the release of lead cations from tableware glaze.

9.1.3 Attack by Acid

Organic acid. Some glazes show greater attack from organic acids than from mineral acids having a lower pH. "Strength" of acid is not the prime factor. Organic anions form complex ions in solution which increases the solubility of glazes under attack.

Chelating agents have the ability to sequester polyvalent cations and their presence in a liquid might be expected to increase the rate of attack.[12] The cations of calcium Ca^{2+}, magnesium Mg^{2+}, and aluminium Al^{3+}, which stabilise the siliceous surface film and aid resistance to further attack, interact with chelating species and are nullified in their protective action. Silicon Si^{4+} forms complexes with catechol which are stable in solution and the unstabilised surface coating succumbs to continuing reaction. Gallic acid and tannic acid behave in a similar manner.

Mineral acid. The mode of attack by acid is similar to attack by water. Compounds of silica are attacked by hydrofluoric acid, although the form in which the silica exists dictates to some extent its resistance. The attack is directed towards a breaking of the siloxane bond Si–O–Si in the glassy structure.

(v) $\equiv Si{-}O{-}Si\equiv + H^+F^- \rightarrow \equiv Si{-}OH + \equiv Si{-}F$

Hydrochloric acid and sulphuric acids are not so aggressive.

Hydrofluoric acid. Hydrofluoric acid reacts strongly with glaze. Acid etching of silicate compositions was developed in the mid-nineteenth century and its present-day utilisation is in the decoration of glaze with "acid–gold". Early decorators covered the glaze with paraffin wax through which patterns were cut. These patterns are now reproduced in screen printed resists.

In the presence of hydrofluoric acid or other fluoride ions, silico-fluoride ions are formed.[13] The fluorides are only slightly soluble and dilute acid is therefore used in association with another mineral acid to dissolve the solid waste of the etching.

(a) $SiO_2+6HF \rightarrow SiF^{2-}_6+2H_2O+2H^+$

(b) $Si–O–Si+H^+F^- \rightarrow SiOH+SiF$

The reaction rate depends upon temperature. As the temperature increases, so does the rate of etch, but losses due to volatilisation act as a counter to an excessive rate. Even acid-resistant glazes are attacked and a mixture of 2 parts hydrochloric acid (38%) and 1 part hydrofluoric acid (40%) gives rapid surface removal from such glazes.

9.2 THE EFFECT OF COMPOSITION

The rate of extraction of ions from any glaze is determined by its composition. "Soft" glazes which mature at low temperatures are often high in alkali and, as might be expected, the rate of release of alkali ions from the glaze increases with increasing alkali content and with increasing ionic radius of the alkali ion. Release rates can be reduced by the inclusion in these soft glazes of one or more polyvalent or divalent elements. The presence in the glaze of either Pb, Ca, Ba, Zn as well as Al,[14] Zr or Ti is beneficial to chemical resistance. Alumino-silicate compositions containing large amounts of rare-earth oxides are very resistant to alkaline attack.[15] Multicomponent glazes and frits are normally very resistant to attack by aqueous solutions except when substantial amounts of silicon are replaced by potassium, when they can be corroded by acids.[18]

It was found[16] that if certain ions were present in the attacking solution, they lowered the rate of attack. Additions of calcium, zinc or aluminium hydroxide to the solution slow the attack and many alkaline washing agents for pottery now contain such additions to give protection.

Early workers[17a, 17b] carried out wide ranging research into the resistance of different glass compositions to attack by water, acid and alkali solutions. Glasses were found to have gradings of durability differing according to the nature of the attacking liquid.

At the time of the original work,[19] it was considered that "good" glass had an acceptable level of homogeneity. Provided that the batch was adequately mixed and the melting was carried out for more than a minimum time and above a minimum temperature, the quenched glass was assumed to be uniform.

Evidence for the presence of two phases in glass and in glaze frits has been forthcoming. Such immiscibility manifests itself in two forms, either as a "two-framework" structure (Fig. 9.1) where both phases are continuous, or as a "droplet" structure, where discrete drops exist in a matrix (Fig. 9.2). This latter form is not common.

Fig. 9.1.

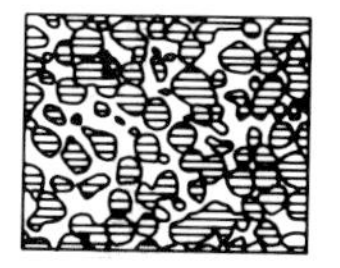

Fig. 9.2.

The durability (however that might be measured) of the "two-framework" structure is determined by the less durable phase, but the durability of the droplet structure is determined largely by the strength of the matrix. During phase separation, changes in composition take place and one phase becomes more durable. If it is that the droplets develop as an unstable, low durability phase, then the durability of the stable phase matrix increases. In the second variety of (droplet) structure, the droplets are formed from the stable, high durability phase and the weakened matrix has a correspondingly low durability. The extent to which this phase separation takes place usually depends on the heat treatment received during the glost or enamel decorating fires, although in some frits the phase separation has occurred spontaneously because of the particular composition. The formation of phases, being dependent upon composition, is influenced even by changes in the source of raw material and by its purity.[20]

Designing glaze compositions which will resist attack by water is not difficult if the silica content is used as a datum. Thermal expansion constraints always limit the level of added alkali and provided the silica content of a glaze is greater than 50% no obvious problems arise from attack by water.

Where there is a risk of water being in permanent contact with glaze (externally mounted electrical porcelain), the silica content of the glaze will be high and the alkali content will be low. It follows that glazes for electrical porcelain are high firing compositions.

Acid-resistant glazes can be formulated by basing the compositions on very high silica contents. Modern once-fire sanitary ware, exposed to alkaline and acid cleaners, have glazes containing about 60% SiO_2. Ceramics for chemical engineering applications have glazes based on 75% SiO_2 contents.

The inclusion of supposedly inert mill components (an opacifer or a pigment) can affect chemical resistance of a glaze by modifying the surface protective layer formed during "leaching". In the glost fire some ions migrate from the mill additions to enter the structure of the glaze. The protective layers formed on the surface during chemical attack are enriched by some ions and contaminated by others. Zirconia has a beneficial effect on resistance and the presence of a reservoir of zirconium ions in the opacifying phase provides a key element for modifying the surface layer.

From a pigment in a glaze, this layer could be enriched with transition metal ions. Of these, for example, copper is positively detrimental in all types of glaze.[21] From a Naples Yellow pigment ("lead antimoniate") both lead and antimony will migrate. In the short time of a glost schedule it is unlikely that any migrating ions could be "latched" firmly into the glaze structure; consequently, they are easily accessible to any leachant; nevertheless, under certain conditions lead/antimony enamels can be very durable.

In recent years much attention has been paid to the leaching effect of lead ions from glaze surfaces. As a component of glaze, lead will not greatly affect alkaline durability, whereas acid durability should decrease with its inclusion.[22]

The history of the use of lead oxide in glaze is a long one and the associated health problems have been known almost as long. Work in the government laboratories[23] in the late nineteenth century pointed the way to good working practice within the factory. Lead should be incorporated in glaze in the most insoluble form, that is as a lead silicate glass. The optimum composition is lead bisilicate.

Many tableware glazes contain lead because of its advantageous effect on refractive index, viscosity, thermal expansion and elasticity, but because of its toxic nature a large number of investigations have been made to understand the mode of leaching of lead from the surfaces of all types of vitreous composition. Crystal glass,[24] low solubility glazes, frits[25] and enamels have all been extensively examined.

Leadless frit compositions utilise the valuable flux boric oxide and many of the most durable compositions contain it, but there are two main restrictions to its unlimited use. On first adding B_2O_3 to frit, the boron ion B^{3+} enters the structure as a glass network-former in fourfold coordination and a more durable, stronger, structure develops. With a continuing increase the point is reached (the point depending upon the composition) where the boron acquires a different bonding arrangement and as it assumes a triangular coordination, the structure becomes weakened and susceptible to attack. There is thus a maximum value for the content of B_2O_3 below which compositions for maximum durability will be found.

Large proportions of B_2O_3 lead to phase separation within the frit. One phase will be high in alkali and therefore more open to attack by aqueous solutions. The value of having alumina and zirconia present to increase the molten viscosity of the frit and to reduce the risk of phase separation should not be ignored.

Zinc oxide has been an important ingredient of chemical resistance glasses and it is used to improve alkali resistance.[26] The addition of calcium oxide to some (sodium) silicate glasses has long been known to be effective[27] in increasing chemical resistance. This effect is similar to the effect of other small divalent ions (Mg and Zn). Barium oxide is often added as a partial replacement for lead oxide or calcium oxide.

Any multivalent ion in the glass surface can retard chemical attack by modifying the surface liquid interface, forming insoluble compounds that do not diffuse into the solution. Ions of zirconium and titanium have low solubilities in aqueous solution; therefore, soon after attack begins, equilibrium conditions are reached.[28a, 28b] Glasses containing high levels of zirconium are particularly resistant to alkali[29] and acid.[30] On this basis zirconia is a valuable component in low melting point glazes with high flux contents, counteracting the weakening effect of large quantities of alkali. Alumina has a similar effect.

The majority of commercial glazes contain some alumina either as an impurity (introduced say from the corrosion of refractories during fritting) or from any clay or as a deliberately added aluminium compound because of its valuable properties. If it was absent many leadless frits could not be waterground during glaze manufacture and low-solubility lead frits would not be so easily made. Alumina is known to increase the alkali resistance[31] of silicate-structured glasses. Small amounts of Al_2O_3 fritted into lead bisilicate are potent at lowering the solubility in acid when the frit is subjected to the standard low solubility test.[32]

9.3 METHODS OF ASSESSMENT

Measurement of definite solubilities of glazes or frits is impossible and the suitability of any composition is best determined by actual service, although the alternative is to expose the glazed ware to conditions imitating those met in service. Such tests would be long term and they lack practicality.

Short-term tests giving data from which a value for a possible life-span can be deduced have a long pedigree.[33] Eventually data from a chosen short-term test will be equated with long-term service-life, but any "durability" or "chemical resistance" value derived from one test in standardised conditions will be different from a value obtained by an alternative method. There is no absolute value.

To utilise the premise of examining the quantity of released ions and so to assess chemical resistance brings with it attendant difficulties. When tableware (leadless) glazes are tested at temperatures above ambient with water, perhaps in an autoclave, some components pass into solution; consequently, it is no longer aqueous attack, but attack from a complex, usually alkaline liquor, containing a number of elements. These elements will not be in the same proportion as they occur in the glaze due to their different solubility factors.

Even though a glaze is made to be homogeneous, because of volatilisation from its surface and solution at the buffer layer, the composition varies across the thickness. It is thus difficult to define the solubility of a glaze because an equilibrium state is not easily definable. What is measured in any control test is a comparative rate of attack and it is usual to grade the results relative to other glazes subjected to similar conditions. Each result is dependent upon experimental conditions and attempts are made to standardise these.

The test temperature is fixed above ambient because test results are needed promptly. However, this artificial difference can give results at variance with the known life span of ware in ordinary conditions.

Corrosion effects depend upon the quality of the glaze surface with highly glossed smooth glazes being more resistant than immature glazes. If phase separation has occurred, the alkali-rich phase will be first attacked. Glazes with a crystalline phase associated with microcracks offer to any reagent an enormous surface area for attack and mobile ions can quickly diffuse into and out of the surface. Crystallisation or devitrification behaviour affects the corrosion process,[34] for example, in the case of lithium disilicate where, in addition to ion exchange and network solution, attack takes place at phase boundaries between crystal and glass.

There are two types of test for assessing the ability of frit or glaze to resist chemical attack. The first type is designed to ensure that the components of the glaze are safe to handle before their application to the ware. The second type is designed to ensure that the glazed ware is satisfactory for its intended use. In both types of test it is important to control the ratio of surface area exposed and the volume of attacking liquid.

9.3.1 Powder Test for Crystal Frit

The resistance to attack by water,[35] acid and alkali can be determined on a representative powder sample. Though designed for glass, the test is applicable to frits.

While it is acknowledged that the exposed surfaces of a glaze powder is not representative of the final glaze surface, an attack on a powder presenting a high surface area to an aggressive liquid is a widely used standardised method.[36] The procedure must be rigidly followed if reliance is to be placed on the results. Increasing the speed of the test by using finer powders increases the variables, because the elimination of bubbles from the wetted powder and the measurement of the reacting surface area become difficult. Increasing the speed of testing by increasing the temperature is an acceptable modification, but because each reacting system is unique the data given by the test must still be related to actual service life.

In the preparation of a powder sample the frit must be free from bubble. Grains are prepared by crushing the glass frit for a minimum of time in a pestle and mortar with vertical impact to reduce the generation of excessive fine dust. From this crushed product a "cut" is taken between two closely-matched sieves, rejecting the fine particles and returning the coarse grains for recrushing. The selected fraction is freed from adhering dust by a gentle wash in an organic solvent (e.g. ether)

Leaching takes place in polyethylene beakers immersed in a thermostatically controlled water bath. After the passage of a prescribed time for the reaction between leachant and glass grain to reach equilibrium, the contents of the beaker are filtered quickly. The filtrate is cooled and made up to constant volume. A simple assessment of durability can be found by titrating the alkali extracted with acid. Alternative leaching rates are assessed by analysing the filtrate for significant ions—silica or lead, for example—which could relate to the level (or absence) of chemical attack.

Exceptionally durable frits can be leached in an autoclave.[37]

9.3.2 Lead Frits

It is obligatory that frits are used to introduce the lead component of a lead silicate glaze. The resistance to acid attack of this lead silicate glass frit will be such that, when ground in a glaze, that glaze conforms to lead solubility regulations. The solubility of a lead frit is expressed as

$$\text{Percentage solubility} = \frac{\text{Weight of lead extracted expressed as PbO}}{\text{Total sample weight of lead compound}} \times 100$$

Much work has been done on the method of assessing the chemical resistance of lead glasses.[38] All methods comprise two parts: the physical preparation and testing of the frit sample and the chemical determination of the leached extract.

In the simplest terms a weighed quantity of dried, powdered, lead frit is exposed to attack by hydrocchloric acid (0.07 M strength) at room temperature. The mixture is filtered, the filtrate made to the mark in a graduated flask and the amount of lead salt in an aliquot of the clear filtrate is determined.

The rate of dissolution depends upon the surface area of frit exposed to the acid, the strength of acid and whether it is agitated in contact with the powder or not, the temperature and the time of the reaction.[39, 40] Some of these factors have a greater effect on the solubilising of lead than do others. The degree of agitation has little effect, but close control of the grain size of the powder is of great importance.[41, 42, 43] Whilst this control is achievable in a laboratory test, some variation in the surface area of the lead frit occurs during the grinding of a lead glaze recipe. Modern regulations require the lead glaze frit to have a "solubility" below the maximum permitted level in the form *in which it is finally used.*

9.4 GLAZED WARE

Most tests are designed to concentrate in as short a time as possible the condition to which the glaze will be exposed in long-term service. Obviously tableware will encounter organic acids and alkaline detergents and tests will include the use of these chemicals.[44] All tests are made on a representative number of samples of production.

9.4.1 Sanitaryware[45]

A number of pieces (about 75mm×25mm×6mm) are taken from a production glazed part of the appliance, one piece remaining untested as a standard. Each of the other pieces is partially immersed in the attacking solution at fixed temperature.

TABLE 9.1. CHEMICAL SOLUTIONS

Name of chemical	Strength of solution	Time	Temperature
	%	hours	°C
Acetic acid	10	16	100
Citric acid	10	16	100
Detergent (Note 1)	(Note 1)	48	60
Hydrochloric acid	(Note 2)	48	15 to 21
Sodium hydroxide	5	0.5	60
Sodium stearate	0.15	48	60
Sulphuric acid	3	16	100

Note 1. This consists of an aqueous solution containing 0.04% (wt/vol) of a condensation product of nonylphenol with 8–10 molecules of ethylene oxide. A suitable solution which contains 0.15% (wt/vol) of the product is obtainable commercially under the trade name "Lissapol N".
Note 2. This solution consists of equal volumes of water and of hydrochloric acid specific gravity 1.18.

After immersion none of the pieces should seem to have lost any reflectivity.

A second series of tests involves placing one spot of each of six liquid chemicals on a sample and allowing them to dry. These chemicals of various strengths are 10% sodium hypochlorite, 3% hydrogen peroxide, amylacetate, $\frac{1}{2}$% solution methylene blue and a 1$\frac{1}{2}$% solution of iodine in methanol. No stain should remain on the glaze, which by implication means there is absolutely no reaction between liquid and glaze. A "resistance to burning" test is also carried out.

9.4.2 Tableware

Lead compounds have been used in glazes from pre-Roman times, although the dangers were not then known. As the evidence grew and new glazes were introduced with greater resistance to attack, methods of rapidly testing and grading new compositions were contrived.

Early subjective tests involved the placing of a pickled onion on the glazed plate and examining the degree of irridesence produced on the glaze overnight. A similar test for vitreous enamel used a slice of lemon.

Eventually glass and pottery quality tests developed around exposure to acetic or citric acid.[46] Some of these tests relate to the assessment of enamel decoration and some are specific to the chemical resistance of glaze. Interaction between glaze and enamel has a profound effect on the levels of cations released during attack and they must be selected together. Much on-glaze enamelling has been replaced by under-glaze decoration.

9.5 METAL ION RELEASE

In order to protect the public from the effects of the release of lead and other toxic ions from ceramic ware designed for the storage, preparation and serving of food for human consumption, many countries require that such ware does not release more than stipulated amounts of designated heavy metals after testing by specified acid leaching methods.

There are many different metal release specifications and moves to unify and simplify them continue. Research organisations[47] have investigated the effect of natural foods on the release of lead from tableware to provide guidance for defining more accurately the working limits of any toxic cations which might be leached out. Throughout the world the duration of the test, temperature, acid composition and the permitted limits of subject cations in the leaching solution vary widely. As a result, the pottery manufacturer is faced with difficulty in knowing the legal requirements for selling into overseas markets.

An international standard has been published, ISO 6486, which is applicable to all ceramic ware likely to be used in contact with food whether the articles are glazed or not. British Standard 4860 is concerned with glazed ceramic ware only.

Almost all the metal release standards in recent years require an extraction test involving the use of 4% $^{v}/_{v}$ acetic acid at room temperature (i.e. at temperatures within the range 19–25°C) for 24 hours. Some countries also require that ceramic ware used in cooking operations (e.g. casseroles) should not release more than stipulated amounts of heavy metals following attack by hot acetic acid for at least part of the test (e.g. BS 4860, Part 2). A few countries demand different tests. For example, Denmark requires a test involving three 30-minute boiling periods with 4% acetic acid with, in addition, a test to assess the acid extraction characteristics from areas of the ware contacted by the lips.

The majority of metal release standards stipulate limits for extractable lead and cadmium only. Occasionally limits for other elements are included, for example Sb, As and Ba. Limits for lead and cadmium differ from standard to standard although those for cadmium are always less than those specified for lead within a particular standard, reflecting the greater toxicity of soluble cadmium ions. Typically cadmium limits are one tenth or less of those for lead.

There are four ways of expressing the release:

1. As the weight of metal released in micrograms (μg).
2. As the weight of metal released per unit area, usually in milligrams per square decimetre (mg dm^{-2}).
3. As the weight of metal released per unit volume, usually in milligrams per litre (mg l^{-1}).
4. As the concentration in the test solution in parts per million (ppm).

Milligrams per litre and parts per million are almost equivalent at the low levels of heavy metal ion which necessarily must not be exceeded to ensure compliance with the specifications introduced (i.e. less than 20 mg l^{-1}).

Recent specifications stipulate different limits for flatware and hollow-ware items and differentiate between different sizes of hollow-ware. The use of mg dm^{-2} for stipulating limits for flatware has generally occurred concurrently with the segregation of flatware and hollow-ware articles, the (square decimetre) area referring to that of the surface exposed to the acid extractant or to the plane enclosed by the rim of the article. The use of mg dm^{-2} for flatware is considered more appropriate than mg l^{-1} due to the large ratio of surface area to fill volume of flatware compared with hollow-ware. Using mg l^{-1}, the highest results would be obtained with the shallowest ware, although such ware is in practice least likely to result in extraction of toxic metals as it is normally used in conjunction with solid rather than liquid foods.

In performing a metal release test on flatware, the volume of acid in the article will greatly affect the magnitude of the result if it is expressed in mg l^{-1} or ppm. Any slight differences in the filling height of an almost filled plate will have a marked effect on the volume of acid employed, although the surface area of the plate covered by acid will be little changed. Expressing the result in milligrams per square decimetre will give more consistent and therefore more meaningful results.

Most regulations require that all pieces tested conform to the stipulated limits although some do allow a mean release level from a number of tests to be taken for certain categories of ware.

British Standard BS 4860, Parts 1 and 2

Statutory Instrument 1975, No. 1241, The Glazed Ceramic Ware (Safety) Regulations, requires that ceramic ware in, or on, which food or drink will be stored, prepared or served complies with BS 4860. Part 1 specifies the test method and permissible limits of metal release from glazed ceramic tableware. Part 2 is concerned with metal release from glazed ceramic cookware.

The conditions laid down in Part 1 are typical of those employed in almost all "cold" tests and are:

Acid extractant	4% $^{v}/_{v}$ acetic acid
Time of extraction	24 hours
Test temperature	19–21°C

As cadmium release has been found in some cases to be influenced by the lighting conditions, the test is required to be carried out in the absence of light, either under an opaque cover or in a dark room.

The test on an article of tableware is carried out by first washing the piece with tap water containing detergent, rinsing with distilled or deionized water and drying either by draining or with a soft clean cloth. Without in any way contaminating the surface of the sample, the article is filled to its rim with 4% acetic acid and covered to prevent contamination and evaporation. The acid is allowed to stand in contact with the ware for 24 hours in darkness.

The cookingware test required by Part 2 involves a similar procedure except that the ware to be tested (including any lid) is heated to 120°C prior to being two-thirds filled with boiling 4% acetic acid and covered (with its own lid if possible) to minimise evaporation. The vessel is maintained at 120°C (±5°C) for 2 hours before removal from the source of heat. It is allowed to stand for a further 22 hours. If significant evaporation occurs during the initial 2 hours of the test, more 4% acetic acid must be added to ensure that the area of contact of the acid is not substantially diminished. At the end of the test, extra acetic acid is added to restore the volume to the original level.

At the end of either test, the cover is removed, the solution stirred to make it homogeneous and a sample removed for analysis. Lead and cadmium are determined by atomic absorption spectrophotometry.

BS 4860, Part 1, stipulates different limits for flatware, (plates and saucers) and two types of hollow-ware, the maximum amounts of lead and cadmium which can be released for each category of ware following the metal release test being

Hollow-ware	1100 ml capacity and above 2.0 mg l^{-1} lead; 0.2 mg l^{-1} cadmium
Hollow-ware	Below 1100 ml capacity 7.0 mg l^{-1} lead; 0.7 mg l^{-1} cadmium
Flatware	20.0 mg l^{-1} lead; 2.0 mg l^{-1} cadmium

BS 4860, Part 2, stipulates limits for cooking ware of 7.0 mg l^{-1} for lead and 0.7 mg l^{-1} for cadmium. For all categories of ware, all samples tested must release no more than the stipulated amounts for them to comply with the Standard.

ISO 6486[48]

International Standard 6486 (Part 1, Test method; Part 2, Limits) describes a test similar to BS 4860, Part 1, but stipulating the following limits for the defined categories of ware:

Small hollow-ware	1100 ml capacity and above	2.5 mg l^{-1} lead; 0.25 mg l^{-1} cadmium
Large hollow-ware	below 1100 ml capacity	5.0 mg l^{-1} lead; 0.50 mg l^{-1} cadmium
Flatware		1.7 mg dm^{-2} lead; 0.17 mg dm^{-2} cadmium

The limits set for flatware are expressed in mg dm^{-2} so requiring both the filling volume and reference surface area to be determined before the results can be expressed in the required units.

9.6 LEACHING BY FOODSTUFFS

It is clearly undesirable for any foodstuff or beverage[49] in contact with glazed ceramic ware to be contaminated by ions leached from the glaze which will affect the taste of the product or otherwise have undesirable side effects.[50]

Elements other than lead and cadmium are also leached from glazed surfaces by acidic food juices and *this* contamination can be detected by an alteration in the taste of the stored product even when the amount of leached metallic ions is very low. For example, copper can be leached from copper-containing glazes by acidic fruit juices, e.g. blackcurrant juice. A level of only 10 ppm copper in the juice gives the product an unpalatable, bitter taste. Release levels of 2–20 ppm copper are typical following 24 hours storage of fruit juice in ceramic ware coated with a leadless copper green glaze. Lead-containing copper glazes tend not to be employed because the combination of PbO and CuO gives rise to excessive lead release, generally to above statutory limits. Even small amounts of copper have a marked synergistic effect on lead ion release from otherwise stable glazes.

Alcoholic beverages, in particular whisky, are sold in attractively designed ceramic bottles, but no leaching of ions from the glaze into the beverage must occur such that the quality of the drink is impaired. Lead glazes are not used and only the most resistant leadless glazes are considered. Care has to be taken to ensure that ions such as Al^{3+}, Ca^{2+}, Mg^{2+} and the alkalis are not released to any extent.

9.7 "DURABILITY"

A series of subjective tests devised by the British Ceramic Research Association can be used. Some are based on a comparison between long-term washing machine tests and laboratory designed tests.

For a complete test, sets of ware are sampled for each liquid, with one item retained (untested) as a standard. The ware is first washed with a degreasing agent. The temperatures and duration chosen are in excess of those likely to be met within everyday practice. Ware which gives acceptable results in these tests will therefore have acceptable lives in service.

Acid test. One set of samples is completely immersed in acetic acid (say 5%) without agitation at room temperature for 16 hours. At the end of the test the samples are rinsed and dried prior to comparison with the standard.

Soap test. For this a thermostatically controlled bath is necessary. A second set of samples is completely immersed in a solution ($\frac{1}{2}$%) of soap flakes, free from inhibitors, for 32 hours at 60°C±2°C. The solution is stirred continuously. After the test the ware is compared with the untreated sample standard. This temperature is well above that tolerable for hand washing.

Washing machine agent. For this test the set of samples is immersed in a (0.5%) solution of washing machine agent for a minimum of 32 hours at 76°C±2°C.

An examination of the quality of the glaze surface gloss and an order of attack is rated on a scale of attack ranging from 0 (no attack) to 7 (no gloss).

Numerical Assessment of Gloss Changes[51]

0. No attack observed.
1. Very slight dulling over large part of surface.
2. Very slight dulling over whole of surface, rather heavier in patches.
3. Noticeable dulling over large part of surface.
4. Noticeable dulling over whole of surface, heavier in patches.
5. Gloss largely gone from part of surface, remainder dulled.
6. Gloss largely gone from all the surface.
7. No gloss left.

The subjective assessment of the surface imperfections is difficult. Any irregularities responsible for the noted differences can be highlighted by using dyes or soft lead-pencils; however, gloss measurement is a useful and accurate method of assessing glaze resistance to chemicals.[52]

Tiles. The test, divided into two parts, is applied only to white or cream-coloured high gloss glazed tiles. Each glazed tile sample is first completely cleaned and degreased. One half of the samples in number are partially immersed in hydrochloric acid (3%) for 7 days. The second half of the samples is partially immersed in a solution of potassium hydroxide (3%) for 7 days. At the end of the set time the glaze surface of the tiles is examined for any signs of attack. For approval there should be no surface deterioration.

REFERENCES

1. Moore, H. *Trans. Brit. Ceram. Soc.,* **55,** 589–600 (1956).
2. Eppler, R. A. and Schweikert, W. F. *Bull. Amer. Ceram. Soc.,* **55,** 277–280 (1976).
3. Paul, A. *Chemistry of Glasses,* p. 121. Chapman & Hall (1982).
4. Rana, M. A. and Douglas, R. W. *Physics and Chemistry of Glasses,* **2,** 179–195 (1961).
5. Charles, R. J. *J. Appl. Phys.* **II,** 1549 (1958).

6a. Beattie, I. R. *J. Soc. Glass Tech.,* **36,** 37N–45N (1952) (Bibliography on glass surface corrosion).

6b. Fletcher, W. W., Keir, E. S., Johnson, P. G. and Slingsby, B. *Glass Technol.,* **3,** 195–200 (1962).

6c. Newton, R. G. *Glass Technol.,* **26,** 21–38 (1985).

7. Wicks, G. G., Mosley, W. C., White, P. G. and Saturday, K. A. *J. Non-Cryst. Solids,* **49,** 413–428 (1982).
8. Dimbleby, V. and Turner, W. E. S. *J. Soc. Glass. Tech.,* **10,** 304–358 (1926).
9. El-Shamy, T., Lewins, J., and Douglas, R. W. *Glass Technol,* **13,** 81–87 (1972).
10. Douglas, R. W. and Isard, J. O. *J. Soc. Glass Tech.,* **33,** 289–335 (1949).
11. Atsushi, Nishina and Masaki Ikeda, *Bull. Amer. Ceram. Soc.,* **62,** 683–685 (1983).
12. Ernsberger, F. M. *J. Amer. Ceram. Soc.,* **42,** 373–385 (1959).
13. Paul, A. *Chemistry of Glasses,* p. 121. Chapman & Hall (1982).
14. Paul, A. and Zaman, M. S. *J. Mater. Sc.,* **13,** 1499–1502 (1978).
15. Akio Makishma and Takajiro Shimohira. *J. Non-cryst. Solids,* **38–39,** 661–666 (1980).
16. Geffcken, W. and Berger, E. *Glastech. Ber.,* **16,** 296–304 (1938).

17a. Peddle, C. J. *J. Soc. Glass, Tech.,* **5,** 195–268 (1921).

17b. Turner, W. E. S. and Dimbleby, V. *J. Soc. Glass Tech.,* **10,** 304–358 (1926).

18. El-Shamy, T. M. *Phys. & Chem. Glasses,* **14,** 1–5 (1973).
19. Turner, W. E. S. and Winks, F. *J. Soc. Glass Tech.* **9,** 389–405 (1925)
20. Kumar, B. and Rindone, G. E. *Physics and Chem. Glasses,* **20,** 148–149 (1979).
21. Fabbi, B., Krajewski, A. and Ravaglioli, A. *Trans. Brit. Ceram. Soc.,* **78,** 87–89 (1979).
22. Paul, A. *Chemistry of Glasses,* pp. 136–137. Chapman & Hall (1982).
23. Thorpe, T. E. The use of lead in manufacture of pottery 1899. Brit. 24 Government Paper 8383–1500/93/1901 wt 32982 DaS4, Brit. Gov. Paper 9264 1500/6/1901 wt 6417 DaS4.

25a. Norris, A. W. *et. al. Trans. Brit. Ceram. Soc.,* **50,** 225–256 (1951).

25b. Yoon, S.-C., Krefft, G. B. and McLaren, M. G. *Bull. Amer. Ceram. Soc.,* **55,** 508–510 (1976).

26. Paul, A. pp. 135–136.
27. Dimbleby, V. and Turner, W. E. S. *J. Soc. Glass Tech.,* **10,** 304–358 (1926).

28a. Paul, A. *J. Mater. Science,* **12,** 2246–2268 (1979).

28b. Majumder, A. J. and Ryder, J. F. *Glass Technol,* **9,** 78–84 (1968).

29. Larner, L. J., Speakman, K. and Majumdar, A. J. *J. Non-cryst. Solids,* **20,** 43–74 (1976).
30. Dimbleby, V. and Turner, W. E. S. *J. Soc. Glass. Tech.,* **10,** 304–358 (1926).
31. Paul, A. and Zaman, M. S. *J. Mater. Science,* **13,** 1399 (1978).
32. Smith, A. N. *Trans. Brit. Ceram. Soc.,* **48,** 69–84 (1949).
33. Turner, W. E. S. *J. Soc. Glass. Tech.,* **1,** 213–222 (1917).
34. McCracken, W. J., Clark, D. E. and Hench, L. L. *Bull. Amer. Ceram. Soc.,* **61,** 1218–1223 (1982).
35. ISO 719.
36. BS 3473 (Revised), Part 3.
37. ISO 720.
38. El-Shamy, T. M. and Taki-Eldin, H. D. *Glass Technol,* **15,** 48–52 (1974).
39. Norris, A. W. and Bennett, H. *Trans. Brit. Ceram. Soc.,* **50,** 225–239 (1951).
40. Norris, A. W. and Vaughan, F. *Ibid.,* **50,** 240–245 (1951).
41. Norris, A. W. *Ibid.,* **50,** 246–256 (1951).
42. Smith, A. N. *Ibid.,* **50,** 257–261 (1951).
43. Recommended procedures for measuring hydrolytic resistance. ISO R719 and ISO R720.
44. Fletcher, W. W., Keir, E. S., Johnson, P. G. and Slingsby, B. *Glass Technol.* **3,** 195–200 (1962).
45. BS 3402: 1969. Specification for quality of vitreous china sanitary appliances.
46. Standard method for acid resistance of Porcelain enamels (Citric acid). ASTM C282–6.

47a. BCRA.

47b. BGIRA.
47c. ILZRO.
47d. Lehman, R. L. Thesis: Acid corrosion mechanism for lead silicate glass (Rutgers University, 1976).
48. ISO 6486, BS 4860.
49. Yoon, S. C., Krefft, G. B. and McLaren, M. G. *Bull. Amer. Ceram. Soc.,* **55,** 508–511, 512 (1976).
50. Bernhard, D. *Ber. Deutsch. Keram. Ges.,* **51,** 169–171 (1974).
51. BCRA permission to quote.
52. Cubbon, R. C. P., Roberts, W. and Salt, S. Members communication by Brit. Ceram. R.A. (1982)
53. BS 1281:1966. Specification for glazed ceramic tiles and tile fittings for internal walls.

Chapter 10
Control Methods

Control of the physical properties of a glaze is a prerequisite of successful, economic, ceramic production by eliminating such manufacturing faults as uneven glaze layers ("short" glazed and "over" glazed ware), fringing, running, dustiness and knocked ware.

10.1 METHODS OF TEST—GRAIN SIZE

Methods are divided into two main groups: firstly, methods which give the distribution of particles by size, and secondly, methods which measure a single parameter from which the operator infers an appropriate distribution of grain sizes.

10.1.1 Dry Sieve Test

Separation by sieving into size fraction[1] is economical and effective for materials with a majority of particles above about 50 μm. The method is easy and popular and reproducible between laboratories if standard sieves are used.[2] For coarse separations of frit batch ingredients dry sieving is used, but when the subject sample contains many fine particles the complete separation into grades is difficult and an air jet technique should be used. In some cases wet sieving of the initially dry sample is used to complete the separation into fractions.[3]

A known weight of dry powder is brushed lightly over the surface of the sieve using a soft camel hair brush until no grains pass through the mesh. The weight retained on the sieve is expressed as a percentage residue of the sample weight. An alternative to brushing is to employ one or more sieves and to vibrate these in a standard controlled manner.

10.1.2 Wet Sieve Test[4]

Glaze slip is conveniently tested whilst still in suspension. The method is fully described in BS 1796.

The presence of large-sized (say 120 μm) particles in a glaze can affect surface texture or by being resistant to solution during the glost fire they appear as perhaps "white speck". It is convenient to test glaze slip for the presence, and proportion, of large particles by wet sieving. The effectiveness of the test depends upon the duration of sieving and the nature of the movement of the sieve. Wetting agents may be added. A known quantity of the test sample is washed successively through a nest of sieves (the finest being at the bottom of the nest) from coarser to finer sieve. A gentle stream of water completes the separation into fractions until the end point is reached when the water flows off practically clear. The residue on each sieve is dried, weighed and reported as a percentage of the sample used.

Electro-formed precision micro-mesh sieves are available for grading suspensions down to 5 μm.[5] Their use is more for research work than for routine control of glaze manufacture.

10.1.3 Particle Size Measurement[6]

At each stage of manufacture, knowledge of the particle size of the reacting components is important. Reactions in the frit furnace depend in part of the mesh size of the batch components. The rate of grinding in the mill is influenced by the size of the feed. The rheology of the glaze slip and the controllability of its application to the ware are inter-related and directly affected by particle size. All reactions in the glost kiln are dependent upon grain size and the manner in which the grains are packed.

To define the particle size of his glaze the technologist is faced with a problem. Ideally a single dimension would be best, and if all grains were spheres this would be possible, but the raw materials of glaze when crushed and ground do not form regular geometrical shapes. Clays are platey, frits and quartz form chonchoidal slivers, felspars are rhomboidal, wollastonite is needle-shaped, whilst pigments can have all types of crystal shape. A compromise is reached. A mixture of particles is defined in terms of particle size distribution, surface area and limiting particle size.

A value for particle size is established by measuring a characteristic property and arbitrarily relating this to a sphere describing the result of the determination as being equivalent to a sphere of diameter *d*. Many methods are available for measuring this property. Different measurement methods *do not* give comparable results because different characteristic properties have been measured. Often it is convenient to choose a method relating to the behaviour of the individual particles in a liquid because most glazes are applied as suspension in water.

Methods[7] of this type are elutriation, sedimentation, turbidity measurement, laser[8] and electronic counting. Results from these methods can be supplemented by those obtained from dry tests such as sieving and microscopical counting. Unfortunately no single method has universal acceptance.

With the components ground to the correct grain size the fluid condition of a slop glaze is determined by several factors. It is over these factors that control must be exercised.

The slop density. The ratio of solids to liquids, normally defined by the weight per unit volume.

The presence (or absence) of a colloid. By inference, this factor is affected by the presence of soluble salts in the water. These might have arisen either from attack on the glaze components, particularly any "borax" frit, or as a deliberate addition of an electrolyte. Adding extra colloid material or treating that which is already present is the principal method of exercising control over the application properties of the glaze.

Sources of colloid material used by glaze technologists are ball clays, bentonites, cellulose derivatives, starch or even superfine glaze.

10.2 SLIP DENSITY

When glaze slips are prepared for dipping, their fluidity can be adjusted to suit the size, the shape and the porosity of the biscuit or the condition of the green clay ware. Very porous substrates require the glaze to have a lower content of dry glaze per unit volume than do denser ceramics, i.e. they have a lower slip density.

Firstly, surplus water is decanted to raise the slip density above that actually required. It is then stirred and sieved to give a uniform consistency. Once the slip density is determined, extra water is added to obtain the desired density. In Imperial units the slip density is quoted in terms of the weight in ounces of one pint volume. In metric units the slip density is quoted in terms of kilograms per litre.

10.3 FLOW TIME[9]

Flow cups provide a simple empirical test for the viscosity of some glaze slips during manufacture. A flow cup is a small container with precise dimensions and drilled with an accurately dimensioned outlet. Variations in viscosity are indicated by differences in flow time. Unless deflocculated, most glazes have some thixotropy; consequently, the flow time is influenced by the amount of mechanical disturbance suffered by the sample before and during the test. Highly thixotropic slips do not register flow times. Flow times should not be converted into absolute viscosity values because of the risk of anomalous results.

10.3.1 British Standard Flow Cup

British Standard 3900 A6 describes the procedure for measuring flow times of liquids, using a standard flow cup. The cup having a capacity of 100 ml is made from a corrosion-resistant material and has an orifice nozzle made, as an insert, from a scratch-resistant alloy. During the test the cup is supported in a frame on a level stand.

With the orifice closed by the finger, the cup is filled until the slip (at 20°C) overflows into the gallery. Any excess slip due to the build up at the meniscus is levelled off with a straight edge. The finger releases the flow and the time for the slip's escape is measured. Alternative flow time measurements are specified.

1. The break point procedure, where the flow from the orifice is timed until the first occurrence of a break in the thread of liquid.
2. The fixed volume procedure, where the flow of a fixed volume (50 ml) into a measuring cylinder placed under the orifice is timed.
   The test can be repeated and an average value taken. If excessive variation in the flow time is recorded, then it is due to the anomalous flow properties of a thixotropic slip.

10.3.2 Seta-Zahn Viscometer

The instrument is designed for the quick measurement of viscosity of many types of liquid on site without the need for accurate test procedural preparation. It consists of a stainless steel cup, capacity 44 ml, which has a precision-measured orifice. At the top of a long handle a small free-moving ring allows the cup to hang vertically. Cups having differently sized orifices are capable of measuring a range of viscosities.

The cup is filled by immersing it completely in the liquid slurry. It is lifted from the slip and the length of time required for the sample to flow through the orifice until the flow breaks is measured. This time is recorded as seconds of flow.

10.3.3 Torsion Viscometer[10]

This simple instrument was designed for rapidly testing Newtonian liquids, but finds wide application for the testing of glazes. It consists of a vertical torsion wire, a flywheel mounted over a scale graduated to 360° with a solid cylindrical bob suspended below the scale. The damping effect of the glaze slip on the movement of the bob after the flywheel has been given a twist of one full revolution is a measure of viscosity. Different diameters for torsion wire and cylinder are available to extend the range of viscosity which can be measured.

Before the test begins the torsion wire is primed with a complete clockwise revolution of the flywheel which is held by a catch. Prior to the measurement, the glaze is thoroughly shaken to destroy any structure which builds up on storage. The glaze is poured quickly into an open cup and brought promptly under the instrument so that the cylindrical "bob" is immersed. The catch on the flyweight is released and the "bob" rotates under the effect of the 360° torsion in the wire.

A pointer on the flywheel moving over a circular scale registers the number of degrees overswing. Depending on whether or not the pointer passes the rest position, the "overswing" is either a positive or negative angle. The test is repeated to confirm this overswing.

With confirmation of the reading registered, the torsion is again added to the wire and the glaze is left standing in the instrument for a further minute. The catch is released and an overswing recorded. The difference between this last reading and the original fluidity overswing recorded as the thixotropy of the glaze.

Values of fluidity so obtained must be considered in conjunction with the slip density.

Viscosity

The property of viscosity is important when glaze slips and the maturing of glazes is being considered.

Viscosity is the measure of the resistance to flow of a liquid, i.e. the ratio of shear stress to shear rate.

$$\text{Viscosity} = \frac{\text{Shear stress}}{\text{Shear rate}} \qquad (1)$$

One rheological model is a rectangular body of liquid of very thin layers superimposed on one another. Assume that the top is movable, then a force (F) acting on an area A will pull sideways on the top. The pulling action is defined as shear stress (τ) equal to F/A. As the top layer moves under shear stress (Fig. 10.1), it pulls the layer directly under it. This in turn pulls the third layer. It is this relay action which is transmitted by drag through the rectangular pile as the base is held stationary on the substrate (Fig. 10.2).

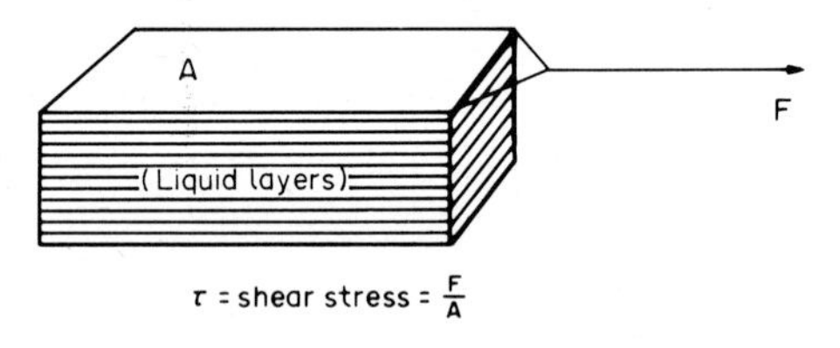

Fig. 10.1

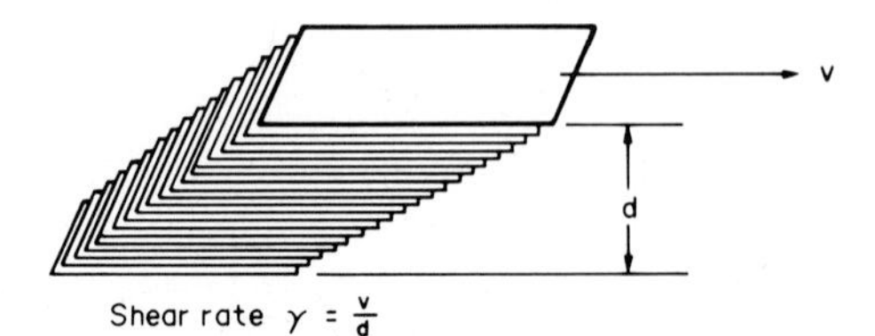

Fig. 10.2

With the velocity of the top layer V and the depth of the liquid d, the velocity gradient is defined as shear rate (γ) and equals V/d.

From equation (1) above

$$\text{Viscosity} = \frac{F/A}{V/d} = \frac{\tau}{\gamma} \qquad (2)$$

Plots of viscosity against shear rate or viscosity profiles are characteristics of particular liquids. These liquids in ceramics can be either aqueous-based suspensions or molten frit. If the system maintains a constant viscosity regardless of shear rate, the flow is said to be Newtonian (Fig. 10.3). Oils are typical of such liquids. Glaze slips which possess thixotropy are not of this type, but those glazes which are deflocculated are approximations of Newtonian liquids.

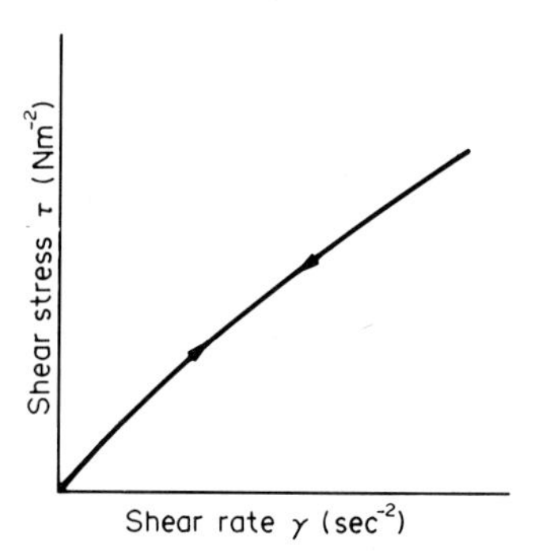

Fig. 10.3. Newtonian behaviour

Molten glasses, free from crystalline inclusions, are considered to be Newtonian.

With a thixotropic liquid there is a range of viscosities that can be measured at a given shear rate. The longer the liquid is subjected to shear, the lower will be the viscosity until a limiting value is reached. The structural loss is temporary. These materials are non-Newtonian but are thixotropic (Fig. 10.4).

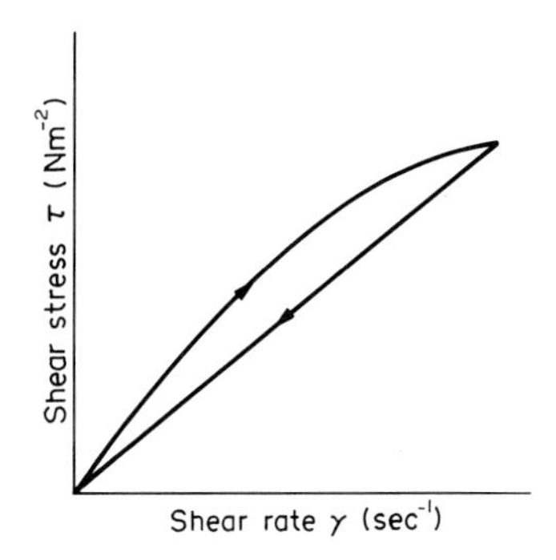

Fig. 10.4. Thixotropy (The area between the curves is a measure of thixotropy)

Many glaze slips are of this type. As the shear stress is released, the viscosity gradually increases, though it might not return to the original value. Glazes which are mechanically applied will have different fluid properties from a glaze for dipping. Their optimum parameters will be determined for the conditions met with at the production speed. Thixotropy is time dependent.

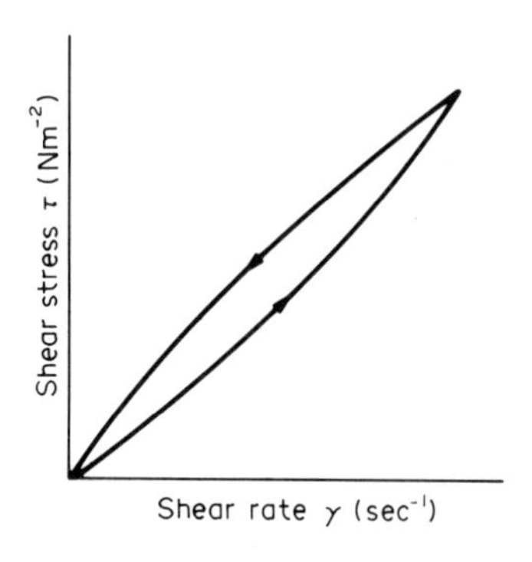

Fig. 10.5. Rheopexy

The reverse of thixotropy is rheopexy (Fig. 10.5) in which viscosity increases with shear, an awkward condition to experience in a glaze slip. It can be rectified by reconstituting the recipe with materials having a greater proportion of very fine particles, $<2.0\ \mu m$ diameter.

Fluidity[11] or Flow Time of Glaze Slip

The degree of movement between particles within the glaze suspension will depend upon the number, size and surface condition of those particles. Finer particles present a great surface area which offers resistance to movement. During the grinding process, and as the surface area increases, there is growing resistance to movement and the fluidity is changed. A glaze made from fine ingredients needs more water to reach the same fluidity as a slip of the same recipe made of coarser components. The effect is very noticeable when clays having different particle size are interchanged in a recipe. Ball clays have a smaller sized particle than china clays, consequently in glazes containing equal amounts of clay, that containing ball clay requires extra water to reach a stated viscosity. Electrostatic forces are responsible for the attraction of one small particle for another, creating larger particles by agglomeration with a consequent effect on fluidity. The drag between the resultant larger particles can be altered by modifying the surface charges with electrolytes. Deflocculated slips have high fluidity and pour easily.

An additive used to modify the fluidity of a glaze (rather than to change slip density) is chosen from a long list of possible materials because of its economy, its insensitivity to small variations, its tolerance to other soluble salts and its stability in the glaze during storage. Some efficient phosphate-based deflocculants "age" and lose their effectiveness with time—an undesirable quality which is emphasised when the glaze is subsequently treated with electrolytes which act as flocculating agents.

Change in fluidity with time. A standard glaze was deflocculated with a sodium polyphosphate and with a low molecular weight acrylic polymer. Fluidities in terms of overswing on a torsion viscometer were followed in the two glazes over a period of 72 hours. (Table 10.1)

TABLE 10.1

	Low mw acrylic polymer	Sodium hexameta phosphate
Original glaze	180°	180°
After addition of deflocculant	320°	340°
After 24 hours ageing	340°	355°
After 48 hours ageing	338°	350°
After 72 hours ageing	335°	340°

This degree of change in viscosity due to ageing is not unusual.

Reaction to calcium chloride. This same glaze in the fully deflocculated state was treated with calcium chloride to flocculate it. The fluidities were again measured over a period of 72 hours as overswing on a torsion viscometer. (Table 10.2)

Table 10.2

	Low mw acrylic polymer	Sodium hexameta phosphate
Original glaze	180°	180°
After addition of deflocculant	320°	340°
After addition of $CaCl_2$ solution	173°	188°
After 24 hours ageing	220°	283°
After 72 hours ageing	205°	310°

The glaze containing the acrylic polymer remained "set up" for 3 days whilst the phosphate-modified glaze gradually "aged," became more fluid and lost its dipping qualities.

Thixotropy

Thixotropy is often apparent in the untreated glaze slip due to the dissolution of flocculating ions from the ingredients.

The property is important in two respects. A glaze containing materials covering a wide range of specific gravity can be stored without preferential sedimentation if the glaze has an adequate degree of thixotropy. Recipes of glazes might be identical, but because either the intended application techniques differ or the porosity of the substrate varies, effective control of thixotropy is necessary. With this achieved, a uniform and consistent thickness of glaze, remains on the newly dipped ware prior to drying. High thixotropy is essential when dipping vitreous ware.

The degree of thixotropy which can be developed in the glaze slip will be dependent upon the characteristics of the glaze. Flocculated glaze is usually thixotropic and electrolytes which increase viscosity also increase thixotropy. Over flocculated glazes will not form a smooth coating even with slips which take a long time to dry on the ware.

Viscosity of Melting Glaze

As the glaze layer reaches peak firing temperature, irregularities of application have to be removed, entrapped bubbles escape and gloss has to develop. At this nominal firing temperature, movement in the glaze must occur to a limited degree. For this to happen the influential properties are the viscosity and the surface tension. Viscosity is the "internal friction" of fluids which creates a resistance to flow. Viscous stresses arise from intermolecular attraction. Surface tension is the property of a liquid surface where it appears to be covered by a thin elastic membrane in a state of tension. There is a tendency for the surface to contract.

The viscosity of many glasses and its relationship with temperature has been studied, but the data refer mainly to compositions not suitable for glaze. In addition, the glasses examined have been homogeneous. Few glazes reach such levels of uniformity.

Between room temperature and "melting" temperature, viscosity changes manyfold. Expressed logarithmically (base 10) the viscosity in this range decreases by a factor of about 20. The steady change is a characteristic of a glass. This relationship between temperature and viscosity is important in the melting and homogenisation of frit and in the maturing of glazes. Optimum melting and removal of bubbles occurs at a viscosity of about 200 dPas.

Changes in composition influence the value of viscosity and its rate of change either by restructuring or by modifying the degree of phase separation.

At low temperatures the delay in the establishment of structural equilibrium becomes large[12] and, where there is a requirement for low glaze firing temperatures, this slowness is compensated for by choosing a composition which includes those elements noted for their effect in lowering viscosity. Viscous glazes restrain the separation into glassy phases and the development of crystalline phases.

The viscosity of a vitreous system in the region of the softening and transformation points is an important property for the fitting of glaze to substrate. The strain which develops due to differences in thermal expansion can be accommodated by viscous flow. How effective this is and its extent will depend upon the constitution of the glaze and upon the rate of temperature change between T_g and M_g, that is, on the rate of change of viscosity. Within this range the viscosity is of the order of 10^8–10^{14} poise.[13]

TABLE 10.3. SUMMARY OF LOW-TEMPERATURE VISCOSITY DATA ON POTTERY GLAZES

(°C)	$\log_{10}\eta$					
Temperature	Lead			Leadless		
	Mean	No.	Range	Mean	No.	Range
600	10.8_5	(7)	9.5_5–11.4	—		—
650	9.6_3	(19)	8.2_5–10.7_5	—		—
700	8.0	(19)	7.2 – 9.1_5	10.1	(10)	9.2 –10.8_5
750	7.4	(10)	7.1 – 8.0_5	8.6	(11)	7.9 – 9.1
800	—		—	7.5	(4)	7.2_5– 7.9_5

At low temperatures the very high viscosity determines the rate of bubble growth. Gases evolved from ingredient interaction and from the porous body will try to expand. Their rate of growth is determined by the glaze viscosity; the rate decreases as the viscosity increases. At the peak kiln temperature the viscosity determines the ease with which these gases can escape, their scars heal and determines the ability of the glaze to flow into a uniform layer.

Viscosity in the temperature range 500–800°C (probably between 10^5–10^{11} poise) can be determined on rectangular bars of glaze[14] firmly held at one end and rotated in a furnace along a horizontal axis. Higher viscosities can be accommodated by modifying the ratio of cross-section to length of specimens.

Above 800°C, the chosen method measures the movement of a loaded bar penetrating a pool of molten glaze, held in a vertical furnace. Problems arise when the glaze reacts with the containing crucible. The results of viscosity measurement on pottery glaze are identical in form with those obtained on commercial glasses.

A rapid assessment of viscosity can be made by noting the flow of frit at high temperature.[15–17] Powdered frit of known particle size is compacted under a standard pressure to form a small cylinder. At first this is fused on a horizontal batt at a known temperature for a fixed time and while still in the furnace the base is inclined at 45°. The glassy button flows a distance depending upon its viscosity and its surface tension at the known temperature.

The alternative method[18] is to assign a temperature at which a flowing glaze or frit pellet reaches a nominated length on an inclined flow block. To have flowed by this standard amount (fixed at 100 mm) the viscosity of each glaze or frit must be equivalent. Using a standard firing schedule, various peak temperatures are chosen spanning that which produces the flow of 100 mm. This flow temperature is calibrated with a true viscosity determined elsewhere.

Reproducibility of results depends upon controlling the important factors of particle size, dimensions and weight of the pellet, heating time and constancy of temperature.

Glaze on a vertical surface will be designed to have a greater viscosity than one for a horizontal surface, thus within the same factory there can be different glazes for flatware and hollow-ware. The more viscous glaze will still mature because of the additional gravitational movement, slight though this might be. For tableware glaze the maximum vertical movement will be less than 1 mm. For some effects the movement might be as much as 25 mm–50 mm.

Thickly applied glazes move more easily than those which are thinly applied and excessive movement can cause displacement in any underglaze decoration unless changes are made to the viscosity promoting elements in the recipe. Glazes which are too fluid are not successful on bodies with high porosity.

REFERENCES

1. Allen, T. Particle Size Measurement 2nd edition, pp. 113–127. Chapman Hall (1975).
2. BS 410.
3. Colon, F. J. *Proc. Soc. Anal. Chem.*, **7,** 163–164 (1970).
4. BS 1796 (1976), Method for test sieving.
5. Daeschner, H. W. *Powder Technology,* **2,** 349–355 (1969).
6. Johnson, R. *Trans. Brit. Ceram. Soc.*, **55,** 237–266 (1956).
7. BS 3406, Methods for the determination of particle size of powders. Part 2: 1963, Liquid sedimentation methods; Part 3: 1963, Air elutriation methods; Part 4: 1963, Optical microscope methods.
8. *Particle Size Analysis,* Ed. N. Stanley-Wood and T. Allen, pp. 425–436. Wiley (1981).
9. British Standard Institution, BS 3900, Part A6 (1971), Determination of flow time.
10. Instrument available from Gallenkamp, Loughborough, LE11 0TR.
11. Dale, A. J. and Francis, M. *Trans. Brit. Ceram. Soc.*, **41,** 178–181 (1942).
12. Douglas, R. W. *Trans. Brit. Ceram. Soc.*, **53,** 748–763 (1954).
13. James, W. and Norris, A. W. *Trans. Brit. Ceram. Soc.*, **55,** 601–629 (1956).
14. Ibid., p. 605.

15. Kinzie, C. J. *J. Amer. Ceram. Soc.,* **15,** 357–360 (1932).
16. Marbaker, A. E. *J. Amer. Ceram. Soc.,* **30,** 354–362 (1947).
17. Dekker, P. *J. Amer. Ceram. Soc.,* **48,** 319–327 (1965).
18. Roberts, G. J., Salt, S., Roberts, W. and Franklin, C. E. L. *Trans. Brit. Ceram. Soc.,* **63,** 563–564 (1964).

Chapter 11
Glaze Additives

Rarely can the ground mixture of frit (if present) and mill ingredients suspended in water be used in a glazing process, straight from the mill with any degree of reproducibility. The rheological properties of the suspension are influenced by the particle size(s) of the several components and these flow properties change with time. Some control must be possible over the thickness of application and its evenness of deposition, consequently additions of rheology modifiers are needed to control slip viscosity, alter thixotropy, overcome sedimentation, improve wetting properties, control drying time and improve the unfired strength of the dry glaze on the chosen substrate.[1] Control over these properties is the means to control the glazing operation. Any addition made to influence one of these properties will often affect others and trials will be needed to find the most suitable combination of materials.

11.1 GLAZE BINDERS

Once applied, and after drying, glaze compositions high in soluble salts or in plastic clay can be handled easily, without damage. However, most dry glaze layers are friable and can suffer damage during the prefire preparation of the ware. These preparatory processes include handling, decoration onto the unfired glaze surface, fettling and placing. The addition of "binders" or "hardeners" to the glaze suspension is consequently necessary. The binder acts as a temporary cement for loosely adherent particles in the unfired glaze. When fully dry, the normally friable glaze surface is thus less easily damaged by any impact[2] or abrasion suffered during later processing. The amount of binder added can range up to about 3% of the dry weight of glaze, with the usual quantity lying in the region of 0.5%. Excessive amounts of hardener can embrittle the dry glaze, yet high levels might be desirable in some decorating processes which involve printing onto unfired glaze.

There are several types of organic binder, usually polymeric, suitable for aqueous glaze systems.[3] The ideal binder burns away freely before 400°C without carbonising, without causing undue shrinkage and without disrupting the cohesion of the glaze layer. Both naturally occurring gums and synthetic polymers are used, sometimes admixed. Glazes modified by these additions are hardened on the surface of the layer. Once this skin is broken there is only a limited degree of binding. Water-soluble organic binders hold strongly to water and the drying behaviour of a glaze is affected by their presence.[4] When a coloured glaze contains hardener, the drying rate might be so protracted that a degree of settlement of the pigment occurs and the fired colour changes by an amount measurable by spectrophotometer. Whether the substrate is clay ware, porous biscuit or is completely vitreous, the thickness of the applied glaze layer can be changed by the chemical composition and quantity of the organic phase in the slip.

The inorganic hardeners, sodium silicate and potassium silicate, have an efficient binding action but have significant side effects on other rheological properties (such as deflocculation). Glazes modified by these materials are hardened throughout the applied layer.

Starch

Starch has wide application in industry as a thickener, extender and adhesive. In glaze it acts as a rheology modifier and improver of mechanical strength in the dry state.[5] There are many grades of starch with different degrees of purity whose major properties depend upon their botanical origin. The amount of ash remaining after combustion is a critical factor when making a choice of starch for use in glaze. Starches can be modified by chemical treatment to alter their physical characteristics. For example, from maize the starch which is extracted has its average molecule size reduced and is converted to the hydroxypropyl derivative. The product is washed before drying to remove those "salts" whose presence in the glaze would affect fluid properties.

Two main types of starch are available. Normally starch powder must be dissolved in hot water to cause it to swell and form a viscous paste, but by suitable pretreatment the starch grains can be made to disperse in cold water. It is usual firstly to make a 10% solution and add that quantity of the thick gum to the glaze to give 0.1–0.5% by weight of starch on dry weight of glaze. The hardening effect is only apparent as a surface effect.

Starch powder should be stored in a cool dry place where there is good ventilation. In aqueous solution, the starch is susceptible to biological degradation and if glaze suspensions containing starch are to be stored for a prolonged period, particularly at relatively high temperatures (*circa* 30°C), a preservative should be added.

Cellulose Ethers

Cellulose is a partly amorphous, partly crystalline solid insoluble in water, but, by etherification, derivatives can be prepared which are water soluble. In these materials the properties of the basic cellulose molecule are modified by the positioning of carboxy groups on the molecule. These products are available in different physical forms which have different handling and dissolving powers. The prime property of carboxymethyl cellulose is the ability to build high viscosities into its solutions and grades can be obtained with low or high viscosities depending on the degree of polymerisation.[6] The binding effect on the glaze increases as the degree of polymerisation increases.

In water the polymerised cellulose swells to give clear solutions whose viscosity depends upon its concentration and the molecular weight of the polymer. The lower viscosity grades are preferred for glaze hardening. The properties of these cellulose solutions can be affected by temperature, pH and the presence or absence of electrolytes or preservatives. Vigorous stirring ensures rapid solution in cold water. It is usual for a 5–10% solution to be prepared firstly from which additions up to 1.0% can be made to the glaze suspension. This is equivalent to 10 g per kilogram of dry glaze. Good quality wallpaper pastes can be used.

Cellulose ethers are chosen as glaze hardeners because their properties are more consistent than are those of natural gums and starches. In a glaze whose degree of flocculation is constant there is an improvement in the stability of viscosity; consequently, layers of glaze whose deposit thickness is controllable can be applied by either brushing, dipping or spraying. Shrinkage of the glaze layer while drying is equally predictable. Not only do they bind the surface of the glaze, but they give a strong bond to the substrate, that is they give "through" hardness.

Choice can be made from a range of cellulose products amongst which are sodium carboxymethyl cellulose, methyl cellulose, hydroxymethyl cellulose and hydroxypropyl cellulose. Some grades are treated to give rapid solution in water although their high surface activity can give rise to foaming. Their solutions are clear, smooth and free from fibre, and for non-ionic grades the ash content is usually very low. Burnout is within the range 250–500°C even in a reducing atmosphere.

Possible side effects on other glaze properties arising from the chosen binder cannot be ignored. Sodium carboxymethyl cellulose, although a preferred binder, acts as a deflocculant in most kinds of glaze. Methyl cellulose is free from alkali and non-ionising and would not so affect the flow of a similar glaze. Hydroxymethyl cellulose solutions are valuable aids in glaze where there is an excess of soluble salt, arising say from the presence of a finely ground borax frit.

Such cellulosic long-chain polymer hardeners should not be added to a ball mill for dispersion in the glaze. Mechanical stress and the heat generated while milling can disrupt the polymer chains and give reduced viscosities. Foaming might also occur. Although some grades are more resistant than others to biodegradation, cellulose solutions for use in glaze require protection against biological and mould attack and from enzyme degradation.

Gums

Natural gums are carbohydrate polymers of high molecular weight. Gum tragacanth is a natural hydrophilic gum obtained from the pith of a bush widely distributed in Asia. For ceramics use, the quality is an off-white or brown coloured translucent powder. Being a natural product it is a mixture of several organic compounds; consequently, gum tragacanth is only partially soluble in water. When mixed with water the gum swells by absorption to form firstly a gel and then, with the addition of more water, it transforms into a smooth semi-transparent sol. These sols have low surface tension and by aiding suspension they are suitable as glaze stabilisers. The viscosity of the sol can be modified by heat and by neutralisation; consequently, it can be used to maintain dense minerals in suspension.

When heated it chars and decomposes without flashing, leaving about 3% of ash. Gum tragacanth is compatible with carboxymethyl cellulose.

Alginates

Alginic acid is an extract from seaweed and two of its salts, sodium and ammonium alginate, can act as protective colloids in glaze. They are soluble in water after prolonged stirring, forming a viscous colloidal solution. At levels between 0.2% and 2% they are effective in improving suspension and increasing the green strength of applied glaze. When heated they combust freely, leaving little ash. Their behaviour is very similar to that of carboxymethyl cellulose solution.

Water-soluble Acrylics

A range of water-soluble acrylic polymers with a choice of viscosities has been synthesised. They are excellent film-formers and are ideal for binding glaze to form a skin strong enough to resist high levels of mechanical impact. Additions average about 1–2% by weight of dry glaze, though the exact amount is best assessed by trial because different grades have different effects on the rheology. Some grades have little influence on the viscosity and others are effective deflocculants. Unlike solutions of natural and artificial gums, the viscosity is unaffected by changes in temperature.

A principal advantage of this type of polymeric glaze hardener is the resistance to biodegradation.

Polyvinyl Alcohol

There is a range of polymerised vinyl alcohol (PVA) compounds which, being soluble in water, act as efficient binders. Low molecular weight PVA dissolves more readily, but considerable stirring is necessary with higher molecular weight polymers to complete the dissolution. PVA does not ionise, but, with increasing concentration and rising viscosity, the solutions tend to gel and in glaze slip these become effective thickening agents. Additions of about 1% to the fluid glaze produce, when dried, tough coherent glaze layers, "through" hardened from glaze surface to body interface. The hardened surface is unaffected by those organic solvents which are used in decorating inks.

Wetting agents help in the dissolution of PVA, but with their use the risk of bubble formation which is already present is heightened. Glycerine can be added as a humectant to toughen the film by retaining a small amount of moisture in the layer of glaze. Starch, dextrine and gelatine act as extenders of PVA and their mixtures can produce very tough glaze on biscuit.

The solutions are prone to bacterial attack and must be protected by the addition of suitable bacteriocides.

Resin Emulsions

Polymerising suitable vinyl or acrylic groups in water either alone or in association with other organic groups produces a range of resin emulsions which can be used as glaze binders. The wet glaze is converted into a tough coherent dry film by the integration of the individual (plasticised) resin particles into a continuous film in which the glaze components are held. This integration depends upon the size of the polymer globules. The film formation is usually rapid as the water in the glaze evaporates. The dry glaze surface cannot resolubilise.

These polymer and copolymer emulsions can be used in association with alginates, gums, polyacrylates and CMC to achieve intermediate properties. Some emulsions have a tendency to foam and their compatibility with antifoam agents must first be assessed.

11.2 ELECTROLYTES

In a glaze suspension there is separation of particles—either single units of coarse ingredients (frit) or agglomerates of the finer colloidal ingredients (clays) by a layer of water. The rate of settling of these components will be different, the smaller grains settling more slowly under the effect of gravity according to Stokes' Law.[7] Colloidal particles might not settle or only very slowly due to the effect of Brownian motion or to an electrical repulsion between particles. With a glaze in storage these different settling rates could lead to significant stratification.

When finely ground material is dispersed in water there is an adsorption of ions on the surface of the constituent particles, one type of ion (either positive or negative) attaching itself to the solid and the other, counter ion, free in the water yet held in close proximity by attraction. As a consequence the solid particles become charged. Each material behaves differently and successive dispersions behave differently.[8] Ion adsorption can be specific to the nature of the fine ground particle by choosing an appropriate electrolyte. Any given solid can be charged either positively or negatively. Only those ions that will combine chemically with the dispersed material are absorbed and the intensity of adsorption is greatest for ions that form the most stable compounds.[9]

When clay is present in the glaze the clay micelle can be charged negatively or positively at will. In this way, by using the pervading clay to affect the whole, the grains of the glaze components can be completely dispersed (deflocculated) or agglomerated into flocs (flocculated). The deflocculating electrolyte increases the electrostatic forces of repulsion on receptive particles so that they move apart. Flocculating salts have the opposite effect. Fine particles no longer repel each other but aggregate into flocs. It is the clay particles that react most readily to flocculants, but other minerals and frit particles react in the same way if those particles are finely ground. The flocs, when formed, settle under gravity.

In these different conditions the flow properties of the suspension will be radically altered. From the moment when, in the mill, new surfaces are formed some reactions occur between the glaze materials and water, and, with ions moving into the water, the electrical charge on the surface of the solid changes. These reactions continue as the glaze ages in storage and the fluidity and thixotropy of the suspension consequently change with time. Experienced dippers can assess the time-related maturity of the glaze by feel alone.

Glazes used in repetitive production cycles need consistent viscosity and thixotropy and these properties are controlled by the addition of electrolytes.

11.3 DEFLOCCULANTS

There are two types of deflocculating electrolytes, the polyanion and the alkali cation.

11.3.1 Polyanion Deflocculants

Soluble complex salts of sodium and phosphoric acid are used as deflocculants. Sodium tripolyphosphate and sodium metaphosphate can be used to assist the dispersion of clays, minerals and pigments in water either during their initial preparation in the mill or later to modify the flow properties of the compounded glaze. The use of deflocculant can affect the hardness of the biscuit glaze surface layer.

Sodium metaphosphate has the ability to form complexes with calcium and magnesium ions, though other ions can also be sequestered. Tetrasodium phosphate disperses efficiently most kinds of inert mineral glaze components. The glassy salts are readily soluble in water; stock solutions for use over a limited time can be made from the powdered salts with gentle stirring in warm water. Very small amounts will reduce the viscosity of pastes to fluid suspensions, the amount needed depending on the ingredients. The slip density of the prepared glaze can be increased easily and therefore the solids content can be modified for any given fluidity or overswing. These phosphates can be added to the mill during the grinding stage to aid the milling of difficult glazes. The effects can be modified by the addition of salts of the alkali metals.

An equally effective range of deflocculants is based on acrylamide or on salts of acrylic acid.[10] Available as liquids, they are instantly miscible with water in all proportions. They can be used to deflocculate already flocculated slips. They are low molecular weight polyelectroytes which can be adsorbed on the surface of the grains in the glaze slip, but unlike the high molecular weight polymers (Section Section 11.4) no bridge building occurs. Instead the charge on the molecule is transferred to the particle and all those having a like charge repel one another.

11.3.2 Alkali Cation Deflocculants

In general terms these are selected from the monovalent alkali salts.

Ammonium hydroxide. When alkali metal cation residues are undesirable in a fired glaze, the weak deflocculant ammonium hydroxide can be used.

Sodium carbonate ("Soda ash"). Though usually added to a slip in conjunction with sodium silicate, sodium carbonate can be used alone. Unfortunately slips as modified age rapidly. Additions are made within the range 0.025–0.15% Na_2CO_3 on dry weight of glaze.

Sodium hydroxide. Not often used despite its strong effect, probably because of its unpleasant nature. The range of addition lies between 0.01% and 0.1% NaOH on dry weight of glaze.

Sodium silicate. This is a widely used reagent, particularly in conjunction with sodium carbonate. This dual chemical system gives close control not only over fluidity but also over thixotropy of the glaze slip. Sodium silicate is economic and readily available in a variety of grades. Amounts of the order of 0.1–0.5% by weight of clay are used. The full effect of its addition to a glaze slip is very dependent upon the quality of the clay used in the glaze. Storage should be closely controlled to avoid deterioration from intereaction with carbon dioxide. Potassium silicate can be used as an alternative.

Sodium oxalate. Sodium oxalate will precipitate unwanted cations in the slip and it is therefore a very active deflocculant. Because of its effectiveness sodium oxalate is sometimes used in particle size analysis.

11.4 FLOCCULANTS

Electrolytes can be added to glaze suspensions to alter the physical properties of the components in water so that the grains no longer repel each other but aggregate into groups called flocs. The action is most readily seen in suspensions of clay whose colloidal particles react differently to various salts. Flocculation occurs when the repelling charge on the particle is reduced to a level which allows mutual attraction forces to operate.

When several particles cohere as flocs they settle more readily than would the individual particles. The sediment which forms is less dense and is often mobile. Some dispersions are unstable and readily flocculate, but others need surface effects to be neutralised or adsorbed salts to be precipitated. The electrolytes which have this effect are divalent cations, acids or salts which ionise with a strong acid radical. Only small quantities are needed, because of their effectiveness, rarely more than 1% being used. This efficiency can be hindered by the presence in the slip of a second, protective, colloid. Gentle agitation can assist in bringing particles together in flocs.

As a general rule the ions leached from glaze components are alkaline and consequently act as deflocculants, "thinning" the glaze to a viscosity below normal application consistency. This effect is one which occurs during storage. Flocculation can reverse this increase in fluidity and there are several effective electrolytes in general use.

11.4.1 Polyelectrolyte Flocculants[13]

High molecular weight polyelectrolytes influence glaze markedly. Quantities added are small, with the range 0.005–0.1%. The important physical constants of these materials are molecular weight and charge.

11.4.2 Divalent Cation or Strong Anion Flocculants

Alum/Aluminium sulphate. These powerful flocculants can be used within the range of additions from 0.01% to 0.25%.

Ammonium chloride[11] gives a stable condition in the short-term, but slips develop an ammoniacal odour during long storage due to interaction with alkali leached from the glaze components. It corrodes glaze-sieving equipment. Excessive use of NH_4Cl causes lead volatisation from the surface of lead glazes at the peak firing temperature.[12]

Calcium chloride as the ability to produce stable flocculation over long periods of storage. It is used extensively at levels of from 0.025% to 0.1%. Adding calcium chloride produces on immediate increase in viscosity.

C lcium hydroxide produces stable flocculation even in glaze slips and some mineral dispersions free from clay. Can be used to flocculate water ground flint suspensions. The optimum addition is 0.025%, but effective flocculation occurs over a wide range from 0.01% to 0.25% depending upon the glaze recipe.

Calcium sulphate. Often the source is potter's plaster, a donor of both calcium Ca^{2+} and sulphate SO_4^{2-} ions. It is best used within the range 0.01–0.25%, though often the amount added is an unknown quantity scraped from a plaster mould. Its effect is longer lasting than calcium chloride.

Hydrochloric acid is used within the range 0.01–0.15% in clay-based systems. In alumina-based systems hydrochloric acid acts as a deflocculant.

Magnesium sulphate. Epsom salt has a long record of successful use in glaze slips. The flocs so formed are stable for long periods.

11.5 SUSPENDING AGENTS

Normally there is a proportion of colloidal material—usually "clay"—in the glaze recipe and this will provide a means of supporting in suspension the inert, non-plastic components. Additional structure-formers, which might be claylike or completely organic, can give increased support and they are included in those glazes where there are constituents—perhaps pigments—with above average particle size.

Bentonite

Because clays in this group have a very small ultimate particle size, less than 1 μm, they are difficult to disperse in water at first, but once this is overcome and water penetrates the clay molecule lattice they form strong gels. Some time must elapse therefore before optimum suspending properties develop to a degree which will prevent the sedimentation of non-plastic mill materials. They are effective at low concentrations of about 0.5–1.0%, and, should any settling occur, it can be readily re-incorporated. At higher concentrations, 1.5–2%, bentonites form thixotropic gels. Fully fritted glazes can be made containing 99% vitreous material, the balance being a bentonite type of clay.

There are many grades of bentonite. Some, like sodium and calcium montmorillonite, have a layered structure; some, like attapulgite and sepiolite, have chain structures. Within the lattice there are cations. In water the clay particles develop a surface electric charge, the magnitude of which depends upon the cations available (Ca^{2+}, Mg^{2+}, Na^{+}, etc.). Clay minerals containing calcium and magnesium ions "swell" and suspensions tend to flocculate. A clay containing sodium ions tends towards deflocculation. The addition of certain compounds to the clay suspension can alter the normal behaviour of the clay through the alteration of the electric charge on the surface. These compounds are electrolytes and they can either thin (deflocculate) or thicken (flocculate) the suspension. However, clays with a high magnesium content have a very low ion exchange capacity and consequently a better tolerance of electrolytes.

Unlike organic thickening agents, clays do not lose viscosity when heated. They do not degrade through bacterial action. The cohesion between clay particles and water can be modified by wetting agents.

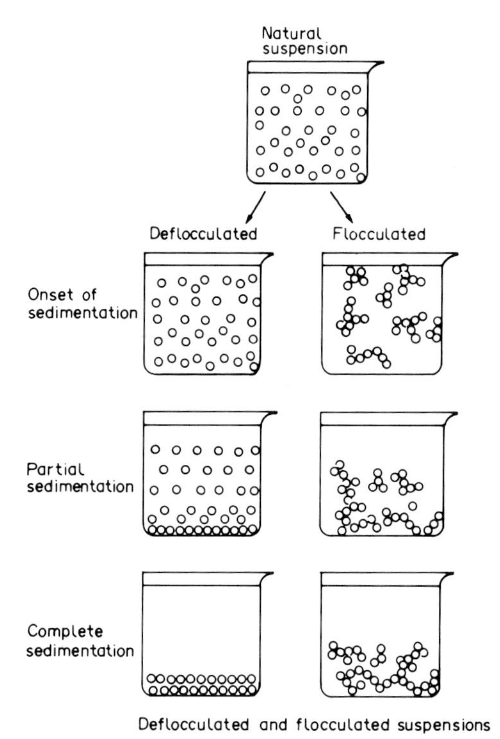

Fig. 11.1. Settling behaviour of suspensions

11.5.1 Artificial Clays

Platey compounds with many of the physical and chemical properties of natural clays, with unusual rheological properties and with the advantage of purity, can be synthesised. These artificial clays based on magnesium silicate resemble the natural clay, hectorite. Suspensions in water are thixotropic and are useful reagents for thickening and for preventing the sedimentation of non-plastic mill additions. Different grades are available which emphasise either the thixotropic property or the level of viscosity and it is possible to impart structure to a system without significant change in viscosity. The interparticle bonds are weak and the structure breaks down easily under shear. This structure rebuilds quickly on standing.

The powders are granular, free flowing and form aqueous dispersions easily in hot or cold water with high shear stirring. Full gel properties develop after a few hours' hydration. Even small quantities confer considerable dry strength to unfired glaze on biscuit or clay because of the compact way their plate-like particles pack.

A secondary function arises because the materials act as buffers in solution, consequently large amounts of electrolytes can be added to the suspension before flocculation or deflocculation occurs. The presence of polyphosphates in the glaze slip increases this tolerance. Such additives are useful where glazes of low fired density are required. The gels formed by these "clays" are stable over a wide range of temperatures.

The partial or total replacement of the normal clay addition by synthetic clay in a glaze improves the fired colour. The slips are stable under normal storage conditions.

11.6 WETTING AGENTS

Certain substances, even in small quantities have the ability to lower the free surface energy of liquids. Such surface active or wetting agents reduce the surface tension of liquid systems and the interfacial tensions of liquid/solid systems. These effects originate from the manner in which molecules of the compound orientate at the interface. A typical molecule has a hydrocarbon (hydrophobic) chain and a polar (hydrophillic) head (Fig. 11.2).

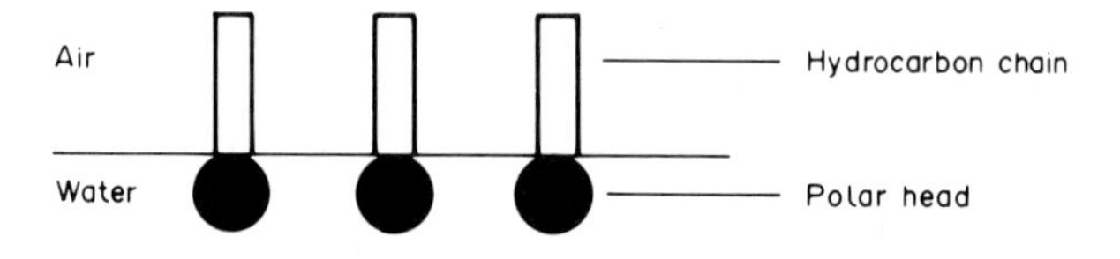

Fig. 11.2.

This orientation varies according to its chemical composition, concentration, temperature and the presence of electrolytes. In water these compounds can dissociate into surface active anions or cations or they can behave in a non-ionic manner and remain undissociated in solution.

Chosen properly, they assist with the dispersion, by stirring, of some glaze components. Without their aid the efficiency of certain underglaze decorating processes would be reduced. In many cases underglaze colour is carried in organic vehicles which are often decomposed in a prefire before glaze application ("hardening on"). The need for this can be avoided. If aqueous suspensions on a solid surface form drops and wetting is poor then the angle of contact, θ, is large (Fig. 11.3). This is the situation with glaze applied on an oily underglaze print.

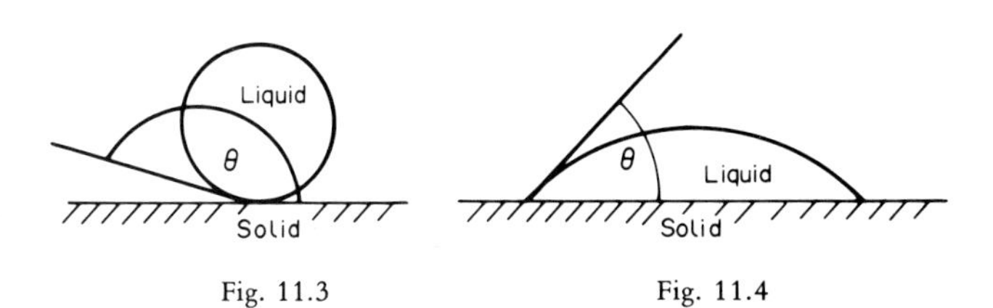

Fig. 11.3 Fig. 11.4

Conversely if wetting is good, that is the surface tension is low, then the angle of contact, θ, is reduced (Fig. 11.4). In ideal conditions this can approach zero. Reduction of the angle of contact is achieved by increasing the surface activity of the water in the glaze slip through the addition of wetting agents. Small quantities of wetting agent in a glaze therefore assist the covering of oil-based decorating inks.

Low- or non-foaming wetting agents are the most suitable for use in conjunction with glaze slips. Adding other types can lead to the formation of foam and bubble, which when trapped in the dipped glaze layer lead to faults in the fired ware (see Section 12.3). Additions to a glaze slip are normally less than 1% on the total slip weight. An excessive amount can result in foaming problems.

11.7 FOAM CONTROL AGENTS

Foam, an aggregate of bubbles separated by thin films or many small bubbles are a feature of some glaze suspensions either directly as a result of the deliberate use of a wetting agent or indirectly as a result of the added organic binder. It is not a desirable condition in glaze because of the association between bubbles and poor glaze condition (Section 12.3). The degree of bubble formation and their permanency depends upon the slip viscosity, pH concentration and the type of binder used. Any factor that favours instability in the film will lead to collapse of the foam. Additions of foam control agents can be made but since, by implication, they must have a degree of insolubility or incompatibility to affect the film surface, the amount used must be small—the application properties of the glaze are not to be affected. Agents with defoaming power at levels of addition of 0.1–0.2% are available.

Specific agents called antifoam agents can be added to prevent foam from forming. They are chemically in the same class as defoamers. The presence of phosphate ion is an effective foam control agent.

11.8 GLAZE SEALERS (FIXATIVES)

Most glazes to which hardeners have been added can receive hand painted decoration. Also direct and indirect mechanically applied decoration can be printed straight onto the unfired glaze. Normal (i.e. hardener-free) glaze has a degree of porosity and, before it will accept transfers or direct screen printing, an impervous surface must be developed by the use of a glaze sealer. As the temperature rises in the early stages of the kiln firing air in the substrate pores expands and its movement would disrupt the decoration without the aid of the sealant.

Sodium silicate additions made to the glaze to act as a binder can produce, at the same time, a smooth surface dense enough to receive transfers. There are indirect ways of treating the biscuit glaze. Polyacrylate emulsions or polyvinyl acetate either as an emulsion or solution are sprayed onto the glaze. After drying they insolubilise to provide an inert surface for receiving decoration by mechanical processes. The chosen sealant must burn out freely with a minimum of carbonising.

11.9 ORGANIC DYES

Within a glazing department, stocks of several compositions can be maintained. Glaze, either raw or fritted, transparent or opacified might look white when ground, whether dried or in suspension. Manufacturers' production numbers might not be a ready identification and organic dyes are used to give indications of different glaze types.

The colouring agents chosen are water soluble basic, cationic dyes with enormous staining powder. Though all now in use are synthetic, they are sometimes referred to as vegetable dyes or stains. During the glost fire the dyes decompose along with other organic material present leaving no permanent effect in the glaze.

11.9.1 Use for "White Glaze"

There are glazing processes where a second application of glaze is needed, and the inclusion of a dye in the subsequently applied glaze can provide a means of product identification and thickness control. After the dipping of vitreous ware, the rims will be more thinly coated than elsewhere and it is practice to compensate for this by applying an extra glaze layer, usually by spray only, to the edges and sharp external angles. An easily recognisable contrast between different glazes is produced by the inclusion of a dye in one recipe.

11.9.2 Use in Coloured Glaze

Despite the presence of ceramic pigment some coloured glazes are white in the ground state. It is a convenient method of identifying stocks of ground (pastel) glaze to add to the slip an organic stain resembling the shade which the glaze will ultimately develop on firing.

The five main colours used are red, blue, yellow, green and violet. Additions of the chosen dye can be made within the range 0.001–0.02% based on the dry weight of glaze. The effectiveness of some dyes is determined by the presence of electrolytes in the slip. Colour changes are possible with changes in pH of the slip.

11.10 GLAZE BREAKDOWN AND PRESERVATIVES

11.10.1 Breakdown of Organic Additives

In storage. The viscosity of solutions of the organic, high molecular weight polymers used as hardeners is dependent upon the chain length and molecular weight of the molecule. If the links in the poly chain are broken then the solution viscosity will be reduced. Such breakages can be initiated by mechanical damage during ball milling, heat, hydrolysis, oxidation and enzyme reactions.

Many natural organic hardeners added to the glaze slip will undergo degradation at a suitable temperature as the result of fungal and bacterial attack. Bacteria can utilise organic compounds as a source of carbon and energy and many can grow in the absence of atmospheric oxygen. Glaze technologists use bacteriostats or bacteriocides to control their growth. Some clays can deactivate organic biocides; therefore, specialised advice should be sought as to the correct grade to use. Yeasts require carbohydrates as nutrient and they can appear where starches and dextrines are used as glaze hardeners. Fungi are small micro-organisms which generally require oxygen for development. They are controlled by the use of fungicides. Microbial growth can lead to the production of an enzyme which will degrade cellulose derivatives.[14] An improvement in resistance to biological degradation has been noted when natural polymer glaze additives are replaced by synthetic organic polymers.[15]

Spoilage of glaze containing organic modifiers can occur with time by the evolution of gas bubble, by the development of malodour and by a deterioration in fluid properties. Were glazes in this poor quality condition to be applied by dipping, then the generated bubbles held within the slurry would remain as discrete entities to be preserved in the vitreous glaze layer as a defect.

During firing. Organic suspending and binding agents as well as the organic mediums used in most decoration processes begin to decompose at temperatures as low as 110°C with the liberation of water and of carbonaceous fumes and gases including carbon dioxide and sundry hydrocarbons. Provided these materials volatilise or distil completely and no solid carbon is formed before the glaze seals, then no bubbles attributable to this source will be formed. The more complex organic polymers and some crude natural additives are more difficult to heat without the disastrous formation of solid carbon. Once formed under these conditions, carbon is difficult to eradicate. Under adverse circumstances carbon will remain as a speck even after reaching the maximum glost temperature to mar the finished glaze.

As a general rule those cellulosic hardeners which are chosen will contain oxygen, preferably joined by a double bond, within the molecule and localised reducing conditions are thereby avoided during the initial stages of the temperature rise.

11.10.2 Preservatives

The dissociation of organic glaze hardeners by bacterial and fungal attack during storage is prevented by adding preservatives.[16] Care must be taken to avoid reagents which are a health hazard to personnel because of their toxicity. It is useful guidance to employ those preservatives which are approved for use in foodstuffs. All major chemical manufacturers can give advice on the choice of safe preservatives.

Many bacteria and fungi acquire resistance to destruction after preservatives have been used for some time, and if continued protection is required then the agent chosen for preservation must be altered from time to time.

REFERENCES

1. Weinand, W. The effect of glaze additives on the processing properties of glazes. *Interceram.* **29,** 290–299 (1980).
2. German, W. L., *Trans. Brit. Ceram. Soc.* **54,** 399–412 (1955).
3. *Handbook of Water Soluble Gums and Resins.* Ed. Davidsson R.. L. McGraw Hill Co. (1980).
4. Smith, T. A. *Trans. Brit. Ceram. Soc.* **61,** 523–549 (1962).
5. Alston, E. Stabilisation and binding of glazes, *Trans. Brit. Ceram. Soc.* **73,** 51–55 (1974).
6. Green H. *Ind. Rheology and Rheological Structures.* John Wiley, New York (1949).
7. Worrall, W. E. *Ceramic Raw Materials.* Inst. of Ceramics Textbook series (Pergamon), contains a comprehensive treatment of this aspect of water suspensions. (1982).
8. Ish-Shalom, M., Yaron, I. and Gans, D. *Bull. Amer. Ceram. Soc.* **50,** 737–741 (1971).
9. Beans, H. T. and Eastlack, H. E. *J. Amer. Chem. Soc.* **37,** 2667–2683 (1915).
10. Alston, E. *Trans. Brit. Ceram. Soc.* **74,** 279–283 (1975).
11. Atkins C. and Francis, M. *Trans. Brit. Ceram. Soc.* **42,** 157–163 (1943).
12. Smith, A. N. *Trans. Brit. Ceram. Soc.* **48,** 85–97 (1949).
13. Galvin, T. J. and Hughes, F. A. Flocculating bauxite suspensions. French Patent 1,470,568 (1967).
14. *Paint Research Bulletin,* Report 6 and 7.
15. Alston, E. Stabilisation and binding of glazes. *Trans. Brit. Ceram. Soc.* **73,** 51–55 (1974).
16. Control lichens, moulds and similar growths. *Building Research Station Digest,* No. 139.

Chapter 12

Glaze Defects and their Cure

To serve its purpose a glaze must be free from faults and a knowledge of the source and cause of imperfections[1] and their elimination from the glaze is essential. For the glaze technologist it is necessary to distinguish between a *glaze* fault and a *glost* fault though both will come within the boundaries of his responsibilities. It is convenient if guidelines are established for a subjective assessment of the quality of glaze surfaces (Table 12.1).

TABLE 12.1 GLAZE APPEARANCE RATING[2]

Rating	Appearance
10	No defects
9	Very fine ripple
8	Fine ripple
7	Ripple and some pinholes
6	Noticeable heavy ripple and pinholes
5	Some eggshelling and pinholes
4	Eggshelling and many pinholes

A glaze fault is serious if it reduces the appeal or value of the article upon which it appears. Nevertheless, an appreciation of the nature of glaze imperfections[2] and knowledge of possible remedial action to avoid their occurrence is necessary. Many faults are manifestations of one or more of unwanted glaze reactions, incomplete glass-forming reactions, adverse kiln conditions, the development of undesirable stresses within the glaze, or poor ceramic practice. Specks which might mar the appearance of a clear glaze could be desirable in a coloured or textured decorative glaze. Devitrification and crazing are often encouraged to occur in textured glazes, but if allowed to develop in a tableware glaze, they would be detrimental and unhygienic. However, other severe faults such as dunting, peeling or cracking have no part to play in functional glazes.

12.1 CRAZING

Crazing is the formation of a network of fine cracks within the glaze,[3] these cracks extending from the glaze–body interface to the surface of the glaze. Often they are concentrated in areas where the glaze is thickest and some cracks might be too fine to detect readily. Unless crazing has been deliberately engendered for artistic effect, as in the case of crackle (or craquelle) glaze, then the cracks detract from the appearance of the ware and reduce the ability of the ceramic to perform its designed function, for one of the principal factors in the definition of a glaze—that it gives an impervious coating to a porous substrate—is thereby negated. The adverse effect crazing has on the acceptability of glazed ware is formally specified.[4,5]

"Instant" and "delayed" crazing are caused by the imposition of tensile stresses on the glaze which are greater than it can withstand. These stresses arise from a mismatch between the contraction of the glaze and the substrate as the temperature decreases below the softening point of the glaze. Glaze can withstand greater compressive forces than it can tensile forces and the technologist therefore designs a glaze to be in slight compression during the normal working conditions of the ware. In general terms, this requires the glaze to have a thermal expansion lower than that of its substrate. This generalisation will be examined in detail.

Different types of crazing will now be discussed.

12.1.1 Primary (instant) Crazing.[6]

Glazed ware, when it is removed from the kiln, is seen to be crazed. If excessive thermal shock is not suspected, then the development of the network of cracks is due to poor glaze fit. The thermal expansion values of either glaze or body, or both, are not balanced (see Section 6.3).

12.1.2 Secondary (delayed) Crazing

Moisture expansion. It was once thought that delayed thermal contraction in the glaze was the main cause of delayed crazing, but once this was refuted[7] the true reason was confirmed as being moisture expansion of the substrate.[8] This expansion reduces the compressive stress in the glaze to a level where the glaze is subjected to tensile stress with the onset of crazing.

Crazing due to moisture expansion might occur days, weeks or even years after manufacture. It is seen in glazes applied to some porous bodies.[9] Water vapour penetrates minor imperfections in the glaze (pinmarks, setting scars, etc.) and is adsorbed within the body pores. This process begins as soon as the ware leaves the kiln and continues inexorably in normal ambient conditions. The extent of this expansion is related to the surface area of the internal pore system in the ceramic. It follows that a truly non-porous body will not suffer from such delayed crazing.

Porosity appears to be a critical factor in the development of moisture expansion, yet, bodies having the same porosity although made of different ingredients, can react to water vapour very differently. The amount and nature of the glassy phase in the porous ceramic influences the magnitude of the moisture expansion effect.[10]

Small additions of talc, dolomite or calcined magnesite (up to about 2%) to an earthenware body will reduce moisture expansion but do not greatly affect porosity. Providing the firing temperature is not excessive, talc has a secondary effect in converting some of the microcrystalline quartz (flint) to cristobalite, the presence of which leads to an increase in the thermal expansion of the body and consequently to an increase in the level of compression in the glaze, so adding to crazing resistance.

12.1.3 Enamel–Kiln Crazing

There are several possible causes of this type of crazing even though the differential expansion, Δ, at 500°C of the glaze and the body has an adequately large value. Enamel–kiln crazing can occur on bodies which have a high cristobalite content. In the early stages of an enamel firing the volume changes associated with a pronounced α–β crystal inversion can reduce the level of glaze compression and a negative stress is induced, causing fracture.

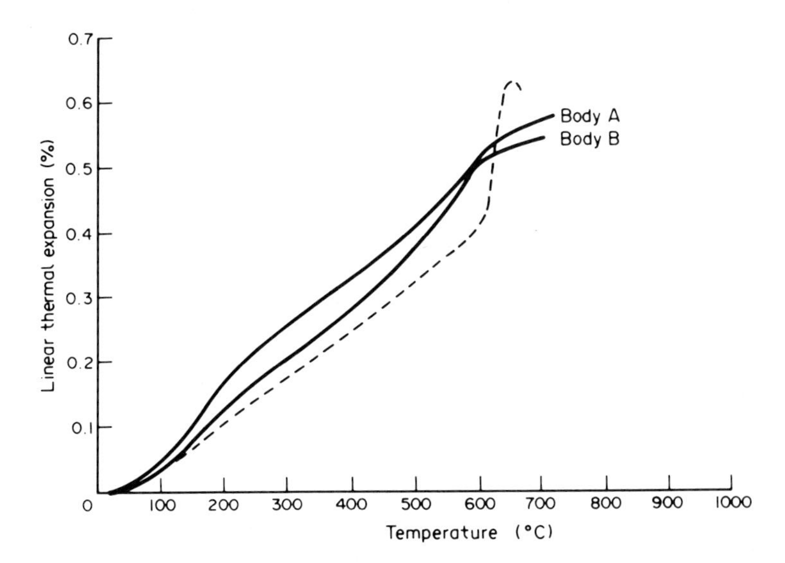

Fig. 12.1

When the thermal expansion curves of body and glaze are compared, the magnitudes of the differential expansion at 500°C can be seen (Fig. 12.1).

The values of the differential expansion, Δ, between a standard glaze and each standard body are not unusual, yet body A has a high risk of crazing in the enamel kiln, whilst body B has a low risk. Rapid passage through the kiln in the early stages emphasises the magnitude of the differential.

With this fault there are often associated lines of small bubbles or "spits" and the cracks in the glaze, if they remain, are usually in concentric rings nearer to the rim than to the centre of the plate. In this case the cause is variable (and severe) thermal shock during the heating up stage of the enamel fire. The edge of the ware quickly becomes hotter, consequently it expands more rapidly than the centre and the ware is under strain. At temperatures below the softening point M_g, the glaze is not soft enough to flow under the resulting stresses and fracture occurs along the stress lines which follow the lines of temperature distribution. The cracks are, therefore, concentric. With an increase in temperature and time these gradients decline and the edges of the cracks rejoin. The craze finally heals when the glaze softens further. It is at this point that air trapped along the lines of the fracture form small bubbles. If the glaze softens even more, these bubbles erupt in a manner which resembles "spit-out".

12.1.4 Salt Cracking

In wall tiles attached to walls, soluble salt migration over the long term from any cement support causes an after-expansion[11] in the body, with the result that the glaze surface would prefer a concave curvature. With the tile fixed to the cement and this curving prevented, the tensile force in the glaze is increased. Eventually the induced tensile force overcomes the strength of the glaze and large radius cracks form.

12.1.5 Cement Pinch Crazing

Size changes in the fixing medium for glazed tiles can have a long-term effect.[12–14] When fixed to a wall, forces other than glaze compression are imposed on the tile. Contraction of any cement backing can give rise to the "cement pinch" effect. The body curves and the stress in the glaze consequently moves from compression to tension to the point where the glaze cracks.

Mastic fixings have reduced faults of this type to a negligible level because, being elastic, they can prevent stresses from being transmitted to the tiles. In addition they do not contribute to soluble salt migration and, because the tiles are fixed dry, the initial moisture expansion is less.

To improve crazing resistance of a free-silica containing glazed earthenware type body, we can make changes.

(a) *To the glaze*

1. Reduce the coefficient of thermal expansion.
2. Increase the softening point.
3. Decrease the thickness of the layer applied to substrate.

In summary, these aims can be achieved by increasing the silica content, the addition of some boric oxide, reducing the level of alkalis, a change in the nature of the alkali, or utilising the mixed base efffect.

(b) *To the substrate*

1. Increase the coefficient of thermal expansion.
2. Increase the biscuit firing temperature.
3. Increase the proportion of silica.
4. More finely grind the quartz or flint.[15]
5. Increase the flux content.

12.2 PEELING

The glaze/body system is mechanically stable under moderate compressive force. Compression is induced in the glaze when on cooling, the contraction of the body exceeds the contraction of the glaze. If the body contraction exceeds that of the glaze by too much and the body cannot resist the stress then "peeling" will occur. The glaze flakes or peels away from the surface of the body, usually along the interface, or where the edges of the fracture cracks overlap adjacent glaze. Peeling will normally appear on rims, edges or on sharply curved surfaces. Since most substances can withstand greater compressive stress than tensile stress, peeling is not encountered as frequently as is crazing. The fault will occur when the compressive force generated in the glaze is greater than the strength of the underlying body.

It is a convenient measure of the degree of fit to note the numerical difference in expansion between body and glaze at a fixed temperature, usually 500°C. Experience suggests that this difference should be in the region of 0.02–0.05%, although the effects of other manufacturing procedures can be critical and this figure should not be taken as absolute. Some glazes, particularly those opacified with zircon or coloured with zircon-based pigments, have greater levels of compression than would be indicated by the thermal expansion differential at 500°C because of the effect these additions have in raising the setting point of the glaze. The consequent fit of the thermal expansion curves is displaced.

Bodies having great strength show less tendency to peel, consequently the fault is rarely encountered on vitreous bodies. This suggests that a solution to the problem of peeling could lie in encouraging a greater development of the glassy bond in the matrix of a porous body by higher firing.

The nature of the glaze–body interface has a marked influence on the effects of glaze compression. During the glost fire of a porous body coated with a highly fluxed low temperature glaze, an overdeveloped interface is formed. The mechanical strength of this layer or of the bond between the three part system, body–interface–glaze, can be low and peeling is initiated even when there is not a large mismatch between the respective coefficients of thermal expansion.

When the fired properties and physical characteristics of the body and glaze are standard, peeling can still occur not because of glaze faults, but because of poorly designed body systems. Migration of soluble salts towards edges and sharply angled surfaces of clay ware on drying produces on the body surface a layer which affects the adhesion of the glaze and so gives rise to peeling. Low level additions of a reactive barium salt to the body will reduce the quantity of soluble salts in the body by precipitating soluble sulphates as barium sulphate. (These soluble sulphates are usually the most troublesome, constituting at least 50% of the total soluble salts present in a body.)

Airborne dust settling on the edges of biscuit ware results in poor glaze adhesion. Sponging water, if not changed frequently, can be responsible for glaze peeling. Inevitably sponging water carries a proportion of fine clay particles and clay ware subsequently sponged with this loaded water is left with a thin, loosely adherent layer of the fine clay. This deposit affects the adhesion between body and glaze. Ware produced by slip casting has its own particular cause of peeling when the microstructure of the cast is faulty.

In some cases the stresses generated by thermal expansion mismatch are great enough to cause complete shattering of the glazed article, giving rise to the term "shivering" as an alternative name for the fault.

To improve resistance to peeling:

1. Increase the thermal expansion of the glaze.
2. Decrease the softening point of the glaze.
3. Decrease the thermal expansion of the body.
4. Reduce the likelihood of the effect of soluble salts by the addition of a soluble barium compound to the body. To achieve this neutralisation add 0.05–0.50% barium carbonate.
5. Renew clay ware sponging water frequently.
6. Store biscuit in a dust free environment and clean ware before glazing.
7. Increase the firing temperature of the body to develop a greater glass content.

12.3 BUBBLES IN FRITS AND GLAZES

Glazes and frits being glassy can retain large amounts of the gases originating during vitrification.[16,17] Some is held in solution and some remains as bubbles of varying size.[18] All glazes contain some bubbles. These bubbles might be few in number and small in size, whilst at the other extreme they might be numerous and have a diameter larger than the average glaze thickness. A glaze on a vertical surface will contain fewer, smaller bubbles than the same glaze would contain if fired on a horizontal surface. Bubbles with different diameters are nominated according to their size and appearance.

TABLE 12.2

Blisters	400–800 μm
Orange peel	200–400 μm
Eggshell	100–200 μm
"Bubble"	80–100 μm
Visible with hand lens	60–80 μm
Barely visible with hand lens	40–60 μm

Many bubbles have little or no effect upon the quality of the glaze surface and are consequently ignored, however, some trade specifications[19] include definitions of what constitutes a gaseous inclusion (blister or pinhole) and place restrictions on their number on any article for sale.

The gases forming the bubbles come from one of the following sources.

12.3.1 From Water in the Glassy Frit

The presence of water in glass has been recognised for sometime though it is not surprising that "water" can be incorporated in glass structures.[20] A model for the structure of water[21] proposed that over small ranges, water molecules were packed in a manner resembling tetrahedral SiO_4 groups in quartz or tridymite according to temperature. Its presence can be detected in glassy frits[22] by infra red spectrometry when various absorption bands can be identified and the amount can be directly determined by thermal outgassing in a vacuum.[23]

The solubility in frits is of the order of 0.1–0.3% whilst concentrations of water up to about 10 wt% can be found in certain geological glasses.

Water is present in the raw materials and in the frit furnace atmosphere, and it dissociates on entering the network to form hydroxyl groups probably bonded to silicon in the glass structure.

$$\equiv Si{-}O{-}Si\equiv + H_2O \rightarrow 2\equiv Si{-}OH$$

Various absorption bands exist and peaks have been found in frits[24] which are characteristic of OH bonds.[25]

The composition of the frit affects the amount of water which can be found in the "crystal" frit; for example, frits have been found to contain a higher proportion of water within their structure when made from alumina hydrate than when prepared from calcined alumina. Subsequent reheating of these frits showed swelling of the melts, a swelling which was attributed to water vapour evolution. The temperature at which the onset of swelling could be recognised was related to the composition of the frit. Bubble can also arise from:–

(a) Air from bubble or foam entrapped in the glaze slip arising either from an excessive addition of wetting agents or from too vigorous mechanical agitation in the dipper's tub. This bubble becomes embedded in the dipped glaze layer.

(b) Air from the voids between the particles of glaze in the unfired layer. In an unfired leadless glaze layer 40% of its volume will be "holes" left by the evaporation of the water from the glaze slip.

(c) From the decomposition and interaction of some glaze components, notably carbonates, fluorides, clays and organic materials. In most cases it is crystalline material in the glaze which is the source of bubble by acting as an anchor for gas or air already present or formed by decomposition.[26,27] Quartz is an efficient anchor for small bubbles.[28] The composition of the gases in small bubbles in glass can be analysed with great precision.[29]

(d) From the decomposition of some constituents due to excessive heat or to their change in valency. Changes in the crystal form of iron oxide can give rise to bubbles in this way.

(e) Water vapour trapped by the rapid sealing of the glaze when excessively fast glost cycles are used. Some water vapour is not released until the temperature is above 500°C when many glazes begin to soften.

The presence of water vapour from any source influences the volatilisation of glaze ingredients, for example B_2O_3, within the maturing glaze.

(f) With porous bodies, water vapour and other gases released from the pore system and entrapped in the glaze during melting.

(g) The breakdown of organic hardeners (see Section 11.1).

(h) The breakdown of glaze contaminants, such as silicon carbide for example.

(i) The breakdown of glaze components by interacting with contaminants. The resultant gases either remain in the mass of the glaze or burst through the surface. Bubbles grow by coalescence when the viscosity of the molten glaze is at an optimum value, but other factors control the size, shape and population of the bubbles in the matured glaze.

The following factors can influence bubble growth and bubble retention.

Thickness of the glaze layer. There is a direct relationship between the diameter of the bubble which forms and the thickness of the glaze layer. In addition, there is a critical thickness of the glaze layer for the degree of gas entrappment.[30] The value of this thickness is directly related to the firing schedule and as this changes so does the critical thickness. Nevertheless, the glaze thickness should be controlled as precisely as possible to an optimum value (See Fig. 12.2).

Surface tension. Surface tension is a controlling factor in the speed at which gloss develops on the glaze surface and, as a general rule, the higher the value of surface tension, then the more brilliant is the gloss. In a system of more than one component there is a concentration at the interface (between glass and

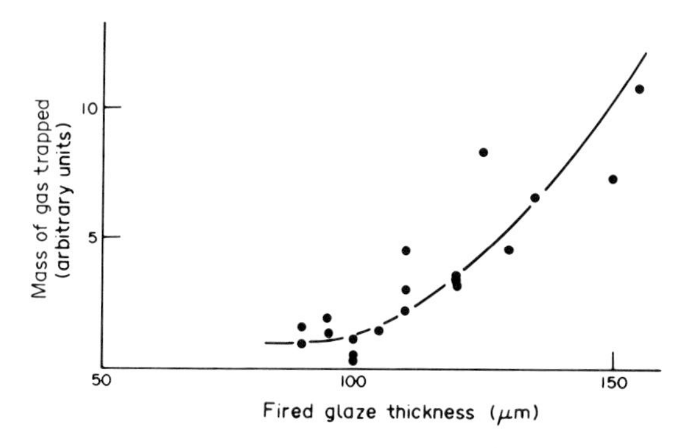

Fig. 12.2. Relationship between glaze thickness and bubble population[30]

atmosphere) of those components which reduce surface tension.[31] As a physical property of glass it is difficult to measure and data are rare. With rising temperature suface tension falls, rapidly encouraging the growth of bubbles.

Viscosity. Within high viscosity systems, bubbles find it difficult to coalesce and grow. The thermofluidity of ceramic glaze means there is a viscosity at which bubbles can begin to grow; the lower this yield point then the easier will bubbles grow. As a viscous glaze matures on its substrate, bubbles will be numerous and small sized. With a fluid glaze the opposite will be the case and bubbles will be fewer and larger. The viscosity of a glaze is controlled by its chemical composition, firing temperature and kiln atmosphere. The relationship between temperature and viscosity is exponential and, therefore, the rate of rise in fluidity with temperature increase is rapid, thus encouraging growth in the size of bubbles.

Glaze composition. Depending upon the glost temperature there are components of the chosen glaze which, if not chosen judiciously, will encourage bubble formation and retention. With high temperature glazes varying the proportions of calcium oxide and magnesium oxide can produce minimal bubble development. In glazes for lower firing temperatures the proportions of alkali metal oxide could exert the same reducing effect on surface tension. Glazes with a single alkali are preferred if the surface tension is to be maintained at a high level. Lead silicate glazes have lower values of surface tensions than alkali and alkaline earth silicates.[32]

Ideally glazes should be fired in a stationary atmosphere so that there develops near to the glaze/air interface a saturation of the volatile components, usually lead and boron oxides. When developing new glazes, firing conditions should equate to the production kiln conditions. "Washing" kiln furniture and the refractories of a new kiln reduces glaze volatilisation. By preventing the loss of oxides of lead, boron and the alkali metals the deterioration of the glaze quality is avoided.

12.4 BLISTERING

Blisters are large bubbles within the glaze which destroy the surface smoothness. Small bubbles are nearly always present in the glaze and they are usually ignored if few in number and below the surface.

If the bubbles burst and sufficient time is not given in the firing cycle for these to heal before the glaze solidifies, the small open craters so formed are referred to as blisters. A large blister is defined[33] as being a raised portion of the glaze surface between 3 mm and 6 mm across. A medium blister is between 1 mm and 3 mm across.

Many factors besides the amount of gas evolved directly from the glaze or indirectly from the substrate can influence blistering. How thickly the glaze layer is applied can determine the degree of blistering. Both the viscosity and the surface tension of the glaze at the point in the firing cycle when bubbles need to be released have a crucial bearing on the permanence of the bubbles.[34] The length of time at this temperature is critical.

Blistering in a glaze can often be prevented or at least the effects ameliorated by following proven guide lines:

1. Reduce the quantity of those constituents which decompose during fritting with the evolution of much gas.
2. Reduce the quantity of those mill additions which release excessive carbon dioxide, sulphur dioxide or water vapour during the glost fire, or which are themselves highly volatile.
3. Completely dry the glazed ware before firing.
4. Reduce the maximum temperature reached in the glost fire. Many examples of blistering are due solely to over firing.
5. As an alternative to 4, the maturing temperature of the glaze might be raised by the addition of either china clay, silica or, in the case of fritted glazes, by the use of harder frits.
6. Adjust the firing cycle to give a slower rate of temperature rise during the early stages of firing, particularly in the stage up to 650°C. This gives the maximum time for the gases to pass through the glaze layer while it is still permeable and prevents the development of a flashed, vitreous, impervious skin on the glaze.
7. Reduce the thickness of the glaze layer. The thinner layer permits an easier passage of any gas and also any entrapped bubbles will be proportionally smaller in diameter.

One particular fault classed as blistering (or sometimes pinholing), shows small craters or bubbles closely grouped together. Depending upon the viscosity of the glaze, i.e. the type of glaze, the holes varying from open scars with sharp ends completely spoiling the finish of the glaze to small depressed bubbles beneath the surface. Examination of the fault by microscope often locates a black speck within the group of bubbles. This contamination can be identified as silicon carbide, SiC, usually arising from an adjacent subsidiary process such as cutting, grinding or polishing. A little silicon carbide is a potent source of bubbles.

When fired in a glaze there is a reaction:

$$SiC+2O_2 \rightarrow SiO_2+CO_2$$

Silica so formed dissolves in the glazes and the carbon dioxide generates a cluster of small bubbles some of which coalesce and burst. Often the silicon carbide will have completely disappeared during the glost fire but from the typical format of the group of bubbles, its presence can be inferred with certainty.

This reaction of silicon carbide is encouraged for some art glaze effects where random surface textures are a necessary part of the design.

If it is essential for grinding or polishing operations to be carried out then alumina-based abrasives should be used. Even scouring pads used to clean ware prior to firing should be suspect. Where silicon carbide wheels must be used then efficient dust extraction systems must be part of the plant and their exhaust should lead only to an outside discharge away from stocks of ware, clay or glaze.

12.5 DIMPLING AND PINHOLING

Dimples and pinholes in glazes are the traces of where bubbles have burst and partially healed. A pinhole is a small hole less than 2 mm maximum dimension.[35] Dimples are usually large blisters which have burst, the glaze has flowed slightly but has not quite healed. Pinholes represent another stage in this sequence, in which blisters have burst and have only just begun to heal. Underfiring or overfiring can be responsible for a dimpled surface. Where the glaze surface is widely marred by numerous pinholes or dimples, the defect is variously called "eggshell" or "orange peel".[36] A glaze which pinholes on one body may not pinhole on another type of ware. There are three key factors, fritting, composition and application. There is no single solution for this fault, since for example either an increase or a decrease in the viscosity of the glaze can be equally effective in suppressing the bubble formation and eruption.

Fritting control. The homogeneity of the frit can be examined for unmelted inclusions, whose source might be coarse raw materials, or result from a too low melting temperature in the frit furance.

Composition. The glaze (or a component frit) might be too soft, becoming fluid early in the firing cycle, and forming a barrier against the escape of gas. A too fluid glaze is often too reactive at the body–glaze interface. Components in the mill recipe are sometimes chosen on the basis of cost and not their suitability. Thus the breakdown of clays might be delayed and the use of different crystal forms of the same material can require different grinding times and can show different speeds of assimilation in the glaze. A stain used in coloured glaze can become unstable at the glost temperature and decompose with the evolution of gas from the small grains of pigment. At peak temperature the glaze might lose volatile constituents.

Application. When glaze not known to produce defects due to bubbles still exhibits a dimpled or pinholed surface, then the technique for applying the glaze must be questioned. Dimpling can be caused in the spray application of glaze either, if the spray gun is too close to the substrate or if excessive atomising pressure is used at the spray nozzle. Pinholing can develop if the glaze application is too thick or if a second coat is applied before the first coat was dry.

Soluble salts. In the preparation of many ceramic stains, a hot water treatment is necessary to remove soluble components. Soluble vanadium salts are a notorious cause of dimples. They readily dissolve in the glaze slip and eventually react to form calcium vanadate crystals which on firing produce a dimple. The remedy is to add a flocculant with which the soluble vanadium forms an insoluble complex in preference to calcium vanadate.[37]

Body. The body must not be too porous otherwise expanded gas erupts through the glaze. Before applying the glaze the surface of the body must be free from dust.

12.6 BITTY GLAZE

Whilst this fault can be detected by touch and the unaided eye, examination of the rough surface with the aid of a good hand lens shows grains of undissolved material in the glaze. The dimensions of these particles are larger than those classified as specks (Section 12.7) and they are greater than the thickness of the glaze layer. The particles disrupt the continuity of the smooth surface. Their identity and therefore their origin can be determined by thin section microscopical examination. Occasionally the frit can be identified as the origin of this fault, but more frequently does the contamination arise from:

(1) a glaze component,

or during

(2) the glaze milling and treatment,

(3) glaze storage,

(4) glaze application.

Origins in glaze components. Some mineral mill additions are known to resist attrition in the ball mill, consequently they are preground to a fine particle size. Unfortunately some mill feed always adheres around the mouth of the mill where it avoids comminution. Good milling practice will avoid this, and provided the glaze milling process includes efficient sieving through fine meshes after grinding, then large grains will be removed.

Origin during treatment. Additions of powdered material (especially clays) to the glaze slip, are difficult to disperse without milling. Waterground materials agglomerate on drying with any soluble salts present, forming a "cement" skin and such aggregates will not breakdown by stirring alone. After additions are made, usually to adjust the fluid properties, the glaze should be resieved and remagnetted.

Origin during storage. Glaze stored in suspension ready for use should be adequately protected from contamination by extraneous dusts carried by draughts or disturbed in the factory environment during works maintenance. All glaze will be sieved after standing in store for even a short time. Even though sensible precautions are taken, bitty-glazed ware can still occur.

Glaze once sieved through a mesh during manufacture may, later, on resieving through the same mesh leave a residue on the mesh. The cycle can be repeated after a delay of several days. Analysis confirms that the residue consists of virtually pure calcium carbonate.

Whilst for practical purposes frits are considered to be insoluble in water, all glasses react in an aqueous environment and they possess a small but measurable solubility. During water-grinding when high temperatures are generated within the mill this rate of solution can be high. The principal ions liberated are sodium, lithium, potassium, calcium and magnesium. With carbon dioxide from the air the reaction to form calcium carbonate is:

$$H_2O+CO_2 \longrightarrow H_2CO_3$$
$$Ca^{2+}+H_2CO_3 \longrightarrow CaCO_3\downarrow+2H^+$$

The reaction takes place slowly and continuously. Calcium ions are released from the frit to replace those which are precipitated as calcium carbonate crystals. Small particles of glaze material or foreign specks act as nucleii for crystallisation. Within the glaze slip the growth is inexorable and uniform, with the particle building layers (as in an onion) to form distinct spheres. At the surface of the glaze suspension or at the sides of the container, the precipitated growth is flakey rather than spherical.

The effects of lime balls can be minimised. All glaze of whatever type or composition should be sieved immediately before use. Commercial glazing units are fitted with sieves through which the slip is circulated. The amount of calcium carbonate abstracted from the glaze composition in this way is negligible and does not noticeably affect the properties. Store rooms or arks for glaze should be selected in cooler parts of the building because the rate of formation of the beads is accelerated by elevated temperatures.

The composition of the component frits can be modified. Leadless frits available to the glaze technologist vary widely in their resistance to attack by water. Recipes can be adjusted to incorporate frits with greater durability. Undermelted, immature frits should not be used.

Providing the glaze slip is not taken to the point of flocculation, small quantities of acid can be added to lower the pH slightly.

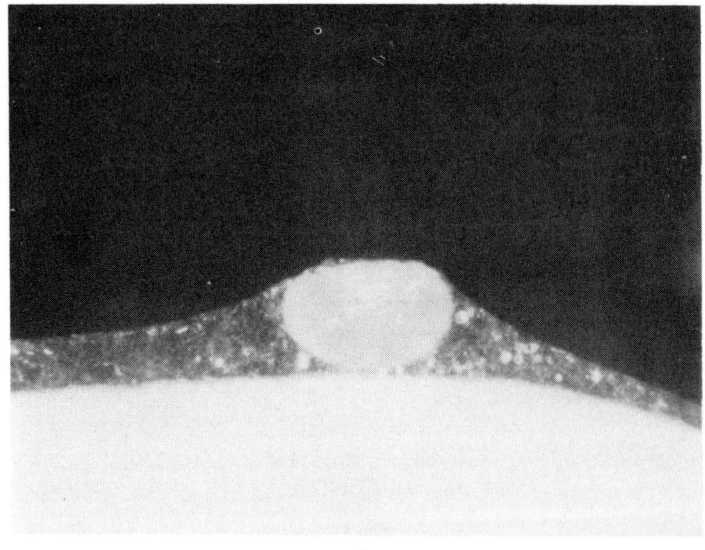

a

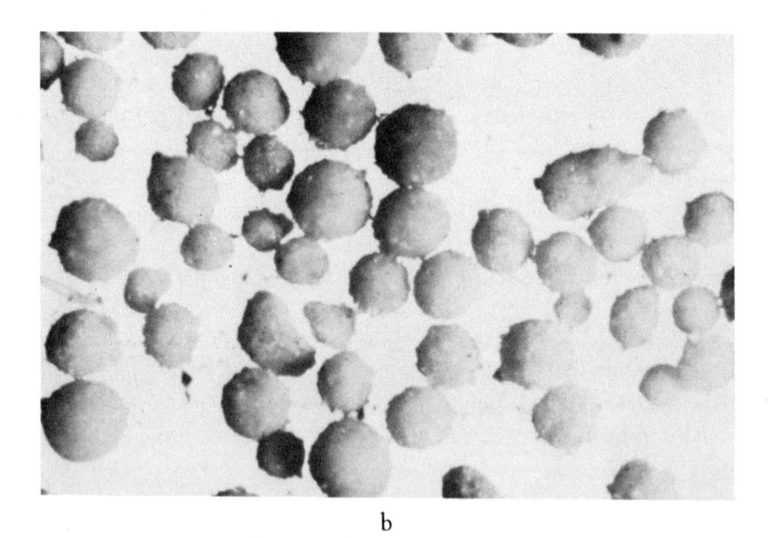

b

Fig. 12.3

a. Section through lime ball in fired glaze (Mag × 30) ... "White dirt"
b. Calcite beads which developed in finely milled glaze, and removed by sieving. Note uniform size. (Oblique illumination Mag × 16)

(These photographs are printed by kind permission of P. C. Robinson and J. D. Royle)

As an alternative, all glazes can be stored ready ground in a dry and powdered condition. A quantity of glaze is prepared appropriate only to that required for immediate needs.

Origin during use. Fragments, of either clay, biscuit or bedding material are washed from the ware during dipping. All ware to be dipped must be cleaned by giving it a light brush, but if the problem persists then the condition of the sieve on the recirculatory system must be suspect.

When the fluid properties of the glaze slip are inappropriate, the thickness of the applied glaze will be inadequate. All fired glazes contain some un-dissolved material, consequently if the weight applied to the substrate is reduced, a point can be reached where the glaze is not thick enough to cover any gross particles. For most purposes when using transparent glazes a minimum fired thickness of 0.15–0.16 mm is recommended. Opaque and coloured tableware glazes have an optimum thickness greater than this at 0.16–0.25 mm. In the glazing of bone china only 0.07–0.08 mm is recommended. For sanitary glazes the average thickness is 0.5 mm. Art glazes range from 0.07–0.25 mm thickness.

12.7 SPECKS

Of the possible faults in ceramic practice, "specking" is probably the most prevalent. A dimension is defined for a "speck" in some regulatory specifications.[38]

Discrete particles of unreacted or unwanted material can be found in virtually all glazed ceramics, but they become classed as a fault only when they detract from the true function of the glaze. Most often specks are dark coloured but an intrusive white particle found in a transparent glaze is a speck, yet this same speck in a white opaque glaze can be ignored. Specks are more evident where the glaze is thickest and where it is overfired.

Specks exist because their constituent materials have shown resistance to solution in frit or glaze and the principles used in the control of glass batch reactions are followed for their elimination. The position of the specks on the ware should be recorded as an aid to diagnosis and as a pointer to the required remedial action. Although there are many causes of specking, the fault is characterised by having its origin in contamination from extraneous or adventitious material, often ferruginous:

(1) in a frit or glaze component,

or arising during

(2) glaze milling and storage,

(3) glaze application,

(4) glost firing.

Origins in the basic ingredients. Raw materials, particularly some clays, contain minerals high in iron, many of which can be removed by magnetting the glaze slip.

It is difficult for the unaided eye to resolve objects of the order of 50 μm and less and the obtrusiveness of many contaminations can be reduced simply by lawning the glaze through a sieve having a similarly sized aperture. However, some small-grained contaminants will partially dissolve during the glost fire to produce a stained area or halo to the speck considerably larger than the original granule. Specks of some pure raw materials remain unmelted as refractory, white inclusions. In a transparent glaze over the white background of its substrate, such specks are often unnoticed.

Particulate organic matter such as occurs in ball clays will combust completely in the glost fire but its presence can be inferred in some coloured glazes by the formation of "white spot".

Where there are differences in specific gravity, the presence of refractory contaminants in raw materials can be confirmed by heavy liquid separation. Most speck-forming contamination will sink in bromoform and the residue can be later examined with a hand lens or microscope to identity the culprits.

Frit rarely contains any strongly coloured contaminants other than iron or carbon, but there might be grains of unmelted raw material. In unground, crystal frit these can be detected if the frit is placed in a petri dish and covered with a liquid with the same refractive index as the frit. For frits without lead the chosen liquid is xylene (refractive index 1.49), and for lead frit the liquid is bromoform (refractive index 1.59). With the refractive indices reasonably close the grains of frit appear quite transparent and any unmelted material is easily seen.

With the frit in the ground form, a sample can be tested for contamination by firing a thick layer of the powder on a small dish through the glost cycle.

Origins during milling and storage. Wear in processing machinery results in particles of potential specks being held in the glaze slip. It is therefore imperative for all glaze making and storage equipment, including sieves and arks, to be checked regularly for defects. Mill media should be regularly examined for signs of fracture and the mill lining should be checked for signs of excessive wear.

Whilst some specks of mineral origin can be eliminated by finer grinding, most will have to be removed by lawning and magnetting, repeating the cycle several times. Aluminium-based metal equipment should not be used for the component parts of glaze processing machinery since any abrasion would produce non-magnetic metallic particles and it is difficult to remove these potent contaminants. In coloured glaze, the addition of coarsely ground or poorly dispersed pigments, being highly refractory compounds, is a frequent source of specks.

"White spot" in opacified glaze is caused by an inadequate dispersion of the opacifier. A variety of white spot seen in transparent lead-free glaze is due to the presence of lime balls (see Section 12.6).

Origins during application. All rules of cleanliness must be followed. Ware, whether in the clay or biscuit state, should be cleaned and brushed before the glaze is applied. There is a risk, where the ware is decorated with an underglaze print, of particles of colour being abraded and released into the glaze slip. Such abrasion can be reduced or avoided by adding binders to the colour or by prefiring ("hardening on") the ware to about 700°C prior to applying the liquid glaze. Foreign particles are known to have been blown through workshop ventilating systems directly onto ware stacked prior to firing.

Detritus from grinding wheels is a notorious source of a variety of specking associated with blistering. This is discussed under "blisters" (Section 12.4).

Origins during the glost fire. The firing system of the kiln can be a source of contamination when dust particles from adjacent ancilliary ceramic processes can be ingested by the re-circulation fans and dispersed within the furnace. Ferruginous dust entrained in the gas feed to burners can be similarly disseminated onto ware in the kiln either when the glaze is still in a porous state, or during vitrification of the glaze.

Powdered mineral thermal insulation can leak between the joints in the kiln structure. Other sources of particulate contamination in this environment are dust abrading from closely set kiln furniture and the splitting or spurting of refractory bits from new saggers or kiln furniture pins and cranks etc.

Droppers associated with continuously-operated tunnel kilns can be classed as a variety of large speck. Some components of the glaze, usually fluxes, volatilise readily in the higher temperature parts of the kiln only to condense on cooler parts of the furnace structure where, being fluxy, they react with the refractory to form a glassy deposit (see Section 12.14).

Droplets from these glassy deposits fall onto the ware below forming round, brownish spots on the glaze surface. Coating the kiln brick work with a special refractory wash (based on fine zircon) is both a preventative and curative treatment although an inadequately adhering wash on kiln refractories and furniture will give rise to intermittent outbreaks of white spot on the glaze.

White spot is more often seen in coloured or opaque glaze with pink glazes being particularly prone. Glazes stained with manganese-based pigment are similarly affected. Organic matter combusts during the glost fire resulting in blistering. A crater forms and into this crater the clear fraction of the surrounding coloured glaze flows (the pigment or opacifier grains being unable to move) to produce the effect of a hole in a coloured background. A longer soaking period at glost temperature can eliminate the fault, though refiring does not always provide a remedy. Slower firing in the early stages will volatilise carbonaceous matter before it is sealed beneath the molten glaze skin.

12.8 CRAWLING

When crawling occurs, irregularly shaped areas that are either unglazed or only partially glazed mar the uniform deposit of fired glaze. The edge of the glaze at the border of the break has drawn back and is smoothly rounded. At this line the glaze deposit is often above average thickness. Crawling can range from small exposed areas of substrate to excessive "beading" where the glaze has formed globular islands of different size, shape and thickness. There are two main causes, one relating to glaze technology and the other to ceramic practice.

Glazes have high but measurable viscosity and surface tension and it is the relationship of these to the substrate which controls the phenomenon of crawling. The majority of commercial glazes do not spread and flow over the ceramic base and crawling is the stable state for most glazes.[39]

Crawling is the result of the lack of wetting or of the absence of bonding to the body by the molten glaze. Any change in the surface conditions of the body which will lead to poor adhesion of the biscuit glaze to the substrate will result in crawling. It is for this reason that dusty ware, ware contaminated by grease or oil, "flashed" or "skinned" biscuit or ware having vitreous casting spots will show a tendency to crawl.

Glazes with a high proportion of clay in their recipe and which are unduly plastic or glazes which have a high proportion of fines produced by excessive grinding crack either on drying or in the early stages of the glost fire. Zinc oxide in a slop glaze held in storage develops a hydrated layer on the surface of each grain causing it to appear to have a larger diameter. These particles shrink significantly when subsequently dehydrated thereby encouraging crack formation. This cracking is evidence of a breakdown in the bond between unfired glaze and body, and these cracks successively widen as the firing cycle progresses and crawling develops. A thickly applied glaze is more likely to crack and crawl. Damage to the unfired glaze by impact or collision can break the fragile mechanical bond between the glaze powder and its substrate. If the impact is excessive, part of the glaze will be detached, then as the temperature increases, surface tension pulls the glaze away from the damaged area.

Gross overfiring can result in the surface of the glaze changing sufficiently to induce crawling.

A special type of crawling is met when glazing "once-fire" ware when the glaze is applied to the clay body. Ware which is then allowed to stand in storage for long periods before firing is prone to crawling. Before the mechanism was understood it was called the "Tuesday crawl". Glazed pieces prepared for weekend firing but not used until the restart of the weekly production were known to be susceptible to crawling. The fault is associated with the re-absorption of water vapour by the powdery glaze layer, causing a slight expansion not matched by any expansion of the dense body. There is a break in adhesion between the two parts which develops into the crawled glaze.

Crawling is associated with the underglaze decorating processes. Some ceramic colour compositions, such as the chrome-based pink and green pigments or cobalt aluminate blue are "dry" and they require an addition of extra flux to encourage wetting of the decoration by the glaze.

The addition of matting agents such as zinc oxide and opacifiers such as zirconium silicate markedly affects viscosity and surface tension and glazes containing these are very vulnerable to the fault of crawling.

Several courses of action can be followed to avoid or to cure the fault.

1. Use a glaze composition only on the substrate for which it is designed.
2. Keep the clay content of the glaze as low as possible and avoid the use wherever possible of large quantities of plastic ball clays. Some clay can be replaced by calcined clay–chamotte or "Molochite".TM
3. Do not overgrind the glaze. Each glaze type has its own optimum particle size range. Zircon opacified glazes are more finely ground than tin oxide glazes and have a higher surface tension. To avoid crawling in all viscous opaque glazes the grinding should be strictly controlled.
4. Keep the quantity of opacifiers to a minimum. If zinc oxide is a necessary ingredient then reduce its activity by precalcination. Components which evolve large amounts of gas, such as colemanite, should be removed. A change in the grade of opacifier is a possible option. Fine grades of zirconium silicate can be replaced by a grade having a larger particle size.[40]
5. Avoid overglazing, keeping the thickness of the layer to the minimum necessary to produce the required effect.
6. Other actions which can be taken when glazing clay ware are to apply the glaze with the clay substrate at zero moisture content and fire the ware as soon as possible afterwards. The effect of soluble salts in the clay body can be reduced by the addition of barium carbonate.

When the fault persists then the effects can be ameliorated by:

1. Slightly wetting the surface of the ware before glaze application.
2. Adding a small quantity of a low foaming wetting agent to the glaze suspension.
3. Adding a small amount of binder or hardener to the glaze (see Section 11.1). Hardeners slow down the rate of drying of the glaze, helping the glaze to wet the substrate. They strengthen the dried glaze layer and so prevent cracking.

12.9 STARVED GLAZE

Glazes which lack shine may do so for several reasons and starved glaze describes a condition where a normally high gloss surface is dull and lacking in reflectivity. The factors which influence the condition are:

1. Glaze thickness.
2. Firing temperature.
3. Kiln setting.

Glaze thickness. When the glaze layer is too thin to cover the normal surface irregularities of the substrate, the texture of the underlying body is reproduced in the surface of the glaze and reflectivity is reduced. The glaze surface is rough to the touch.

Firing temperatures. Selecting a temperature too low or a firing time which is too short for the glaze to mature, gives substandard gloss. On the other hand firing a glaze at too high a temperature, or for too long, is equally deleterious. In overfiring, some volatile components of the glaze system (lead oxide, sodium oxide or boric oxide) can be lost and since these components are usually added for their fluxing power, the overfired glaze lacks brilliance.[41]

Kiln setting. When placing glazed ware near to new or porous kiln furniture or refractories or when firing a mixed load of biscuit and glazed ware there is a risk that the porous structures will absorb glaze vapour, leaving the surface of the glazed ware deficient in the elements needed to give a good gloss and with a silica-rich layer at the outside face. The glazes lose B_2O_3, Na_2O, K_2O and PbO. The absorbancy of the kiln furniture diminishes with successive firings as the pores fill with glaze vapour. Ware with irregularly sited areas of dullness on the glaze is referred to as "sucked" ware.

The recommended remedial actions are:

1. Alter the rheological properties of the glaze either by increasing the slop density or by increasing the thixotropy to ensure an adequate thickness of glaze is deposited on the ware.
2. Choose a firing schedule appropriate to the composition of the glaze.
3. Coat new kiln furniture with a bat wash. Such furniture, unless it is vitreous, is prone to give sucked ware. It can be coated with a suitable batwash to seal the surface and so prevent rapid removal of glaze vapours from the kiln by absorption. Some commercial washes contain a balanced combination of refractory and flux materials.

12.10 SULPHURING (STARRING AND FEATHERING)

A once common[42] but now rare fault, sulphuring is a dulling of high gloss glazes caused by the presence of crystal clusters on the surface of the glaze. These crystals develop from attack on loosely bonded cations, such as lead and calcium, in the molten glaze surface by sulphur trioxide at the end of the firing cycle or by sulphuric acid which condenses on the ware in the early stages of firing. The presence of water vapour and oxides of sulphur together is therefore best avoided.[43] The critical temperatures lie between 600–900°C. Below 600°C, reaction still occurs but at a very low rate and above 900°C, the glazes are too fluid to allow a build-up of sulphates on the surface. The circumstances are similar to those needed for the devitrification of glaze. Many names have been given to this fault, amongst which are starring and feathering which highlight the distinctive features of the marred glaze. However, not all such defects are directly attributable to sulphur.

In the most serious case, crystals of complex sulphates of lead, barium, calcium and magnesium are formed; invariably they are needle shaped and when aggregated in clusters they resemble stars and feathers. Rarely can the dullness associated with these crystals be easily rubbed or polished away. Many sulphates do not readily decompose, consequently sulphured ware cannot be reclaimed by simple refiring and it is necessary to apply a fresh, but thin, coating of glaze over the faulty area before refiring in a clean atmosphere.

The proportions of sulphur dioxide and sulphur trioxide which might be present in the furnace depends on their source materials and their quantity, the amount of free oxygen and the temperature. The fault was common in coal-fired intermittent kilns but with the introduction of cleaner fuels and the use of muffle kilns it has largely disappeared. Sources of sulphur still exist in the furnace atmosphere, in some ceramic body materials and in some grades of fuel oils. To avoid the fault:

1. Do not use fuels high in sulphur.
2. Fire where possible in muffle kilns and ensure a high degree of ventilation within the firing chamber. Sulphur gases can be best removed from the kiln atmosphere before the cooling cycle begins. The extra ventilation not only clears the gases but also increases the cooling rate at the critical temperatures for the sulphur/glaze reaction.
3. Use a refractory wash on kiln surfaces which will preferentially react with sulphur gases.

12.11 CUTLERY MARKING

Glaze cannot be "cut" because of its brittleness, but with hard metals it can be scratched. Any metal cutlery will damage a glaze surface to some degree, perhaps by scratching alone, but on some occasions there is left behind a trace of metal on the glaze. The fault might appear only after the ware has been in service for some time, or it may be incipient as the ware leaves the kiln. Not all glazes will exhibit the fault, a fact which suggests that the composition of the glaze is influential. Some scratches on the face of flatware are due solely to sliding contact with the underside of other ware stacked upon it. This is called "scuffing".

With a rubbing contact between two glazes, glaze and ceramic body or between glaze and metal, damage occurs to the gloss either as an abrasion or as a scratch.[44] The scratch is crossed by a series of conchoidal fractures at right angles to the path. The severity of this damage will depend upon the "hardness" of the source of the damage, the pressure exerted and the relative speed. "Hardness" could refer to the Brinell hardness of a metal or the refractoriness of a glaze.

Under pressure and at high speed the point of contract of one glass surface drawn over another glass surface becomes so heated that light is visible in the dark.

Low power magnification (such as a hand lens) will disclose, at the site of the scratch, a line of chittering where the surface has shattered or ruptured.[45] In the scratch, minute particles of glassy material which have broken away during the formation of the damaged streak can become wedged in the flaws. Many scratches only become evident when the glaze has been later attacked or etched by a leaching ("washing-up") solution.

At low loads, the scratch might be invisible but high loads on the scratching metal edge cause incipient conchoidal fractures. These become more clearly defined as the load increases to a maximum. Conversely, with less pressure the width of the fracture ribs decreases and the distance between the arcs of the conchoidal cracks becomes greater[46] until at very low loadings they do not occur.

Increasing the rate of scratching increases the efficiency of the scratch.[47]

But composition is not the only factor in the appearance of the fault. Both the make-up of the glaze and the furnace conditions[48] during the whole of the firing cycle can have an effect.

Newly produced glazed surfaces (straight from the kiln) and glazes with a very high gloss are prone to scratch more easily. Some zircon opacified glazes are susceptible where adversely large zircon particles are present at or near the surface. Reduction in particle size of the opacifier and of the other glaze components to encourage greater solution of the zircon is remedial, alternatively the zircon can be added in the fritted form. Erosion of glassy material in service by the action of strong detergent to expose a zircon grain can lead to the fault in use.

The ability of some glazes to react with sulphur-bearing[49] atmospheres in the glost kiln or in the enamel kiln has been noted (cf. "sulphuring"). Whilst many kilns use either electricity or low sulphur fuels, some examples exist of obvious interaction with sulphur so the source must be found elsewhere. In a notable example the origin of the sulphur was in the organic base of the decorating gold whose vapours, when trapped in the kiln, reacted with calcium and lead ions in the glaze surface to form calcium sulphate, $CaSO_4$ and lead sulphate, $PbSO_4$. These crystals form a surface sufficiently abrasive to cause metal scratching.

Increasing the ventilation in the kiln and opening the setting are obvious preventive measures.

12.11.1 Silver Marking

This version of the damage to glaze surfaces by metal marking occurs only after the glaze has received an enamel decorating fire.

There is little evidence of the cause of the eventual fault when the glaze is first examined as it leaves the kiln, but after a time in service the glaze will abrade metal from soft types of cutlery.

From the vapours of the volatile fluxes within the kiln, droplets of high lead glass condense on the glazed surfaces. At these comparatively low temperatures (750–850°C) there is a minimum of reaction between the condensate and the glaze. The droplets are too distant from each other to coalesce so there are many small areas over the face of the ware. When the ware is washed during its service the low durability lead glass droplets offer no resistance to any alkaline or acid cleaners and they easily dissolve, leaving shallow depressions. Drawing soft metal cutlery across these depressions produces in them a thin smear of "silver". It is not easy to remove these traces by any means other than by a chemical method.

An old subjective test to assess the likelihood of the glaze developing the fault during its life involved packing enamelled glost plates in straw for one month and, after washing the ware, testing for metal marking with silver wire drawn across the surface. It is assumed the attack on the glaze originated in the sulphur dioxide fumigant in the straw and the modern test obviously developed from this. Sample plates are half immersed in dilute hydrochloric acid for one minute. The ware is washed normally and after drying the plate, the abrasiveness of the glaze surface is tested with silver wire.

The fault can be cured by ventilating the kiln or reducing the density of packing in the kiln.

12.12 CUT AND KNOCKED GLAZE

Although paired, these similarly looking faults have separate causes arising during the application and drying sequence and preventative action is taken by the glaze technologist. Where a glaze is "cut", spots or patches are either bare or only lightly covered. There is a resemblance to crawling on a small scale. Where the fault is "knocked" glaze, there has been an impact and some glaze has been removed from the surface of the substrate.

When glaze is cut, the condition has developed through the inability of some areas of the ceramic body to take up an adequate coating of glaze. With porous bodies any circumstance which reduces porosity, such as contamination of the body surface by dust, oil, water, soluble salts or perspiration, will cause the fault. With vitreous bodies these same contaminants interfere with a uniform application of glaze. In the case of cast ware the formation of a casting spot will cause cut glaze. (The casting spot is a vitreous patch on the surface of slip cast ware at the point where the slip first impinges on the plaster mould.)

Knocked glaze is a result of physical damage to the unfired glaze coating whereupon a flake of glaze is dislodged or an area is rubbed free of glaze. A critical point in the glazed ware production sequence where mechanical damage occurs is when stacking the unfired glazed ware, particularly if the article is imperfectly dried.

To alleviate the problem extra clay can be added to the glaze not only to increase the ability to cover uneven substrates but also to add mechanical strength to the unfired glaze. Further strengthening of the unfired glaze can be developed by the additions of a suitable organic binder (see Section 11.1).

12.13 SPIT-OUT

Spit-out is a fault which can occur during the enamel firing of on-glaze decoration to temperatures in the region of 750–850°C. After such firing the fault, if present, manifests itself by the surface of the glaze containing many small craters. These craters have their origin in bubbles which burst at the glaze surface. The glaze at these temperatures has a viscosity low enough to allow the bubbles to develop *and* burst, but not so low that the scars can heal. In the severest form, the enamel fired glaze has a sandpaper appearance.

Although a glaze fault called china spit is known, the root cause of normal spit-out is almost always the nature of the ceramic body and not the glaze. The common fault is normally confined to earthenware and it has been shown that it is the combination of porosity and free crystalline silica in the fired body which provides a mechanism for the fault.[50] Porous bodies which do not contain crystalline silica rarely show the fault. Underfired vitreous bodies containing crystalline silica can give spit-out if they have porosities above 1%. In one example porous bone china ware had "spit".

Glazed earthenware readily adsorbs water vapour from the atmosphere, during storage, through flaws in the glaze (such as the scars from setting or from sorting), the amount adsorbed increasing as the relative humidity rises and as storage times lengthen. If this ware is later decorated with on-glaze colours and fired in an enamel kiln at about 750°C, desorption takes place through the body's interconnected pore system. This water vapour can be expelled from the pores via "body-connected" bubbles in the glaze to give spit-out.

All glazes, however well fired, contain bubbles. In glazed earthenware these bubbles become connected with the pore system of the body during the cooling stages of the glost cycle as a result of the stresses caused by the crystalline inversions of silica particles at the glaze–body interface. Microcracks between bubbles and silica particles form during the β–α quartz conversion at 573°C and the β–α cristobalite conversion at approximately 200°C when the crystal changes are associated with volume changes. These temperatures are below the transformation temperatures of most glazes and the glazes are rigid and unable to accommodate volume changes. The consequent stresses cause microcracking (see Fig. 12.4) in and around the grains.

Although the source of the fault lies in the nature of the body composition, the incidence of spit-out can be controlled to a degree by the nature of the chosen glaze, because the viscosity of the glaze at enamel firing temperatures is a prime factor in determining the degree of bubble enlargement. "Stiff" glazes will minimise the severity of spit-out because bubbles will find it more difficult to grow in glazes with high viscosities. However, it is now common practice to drill

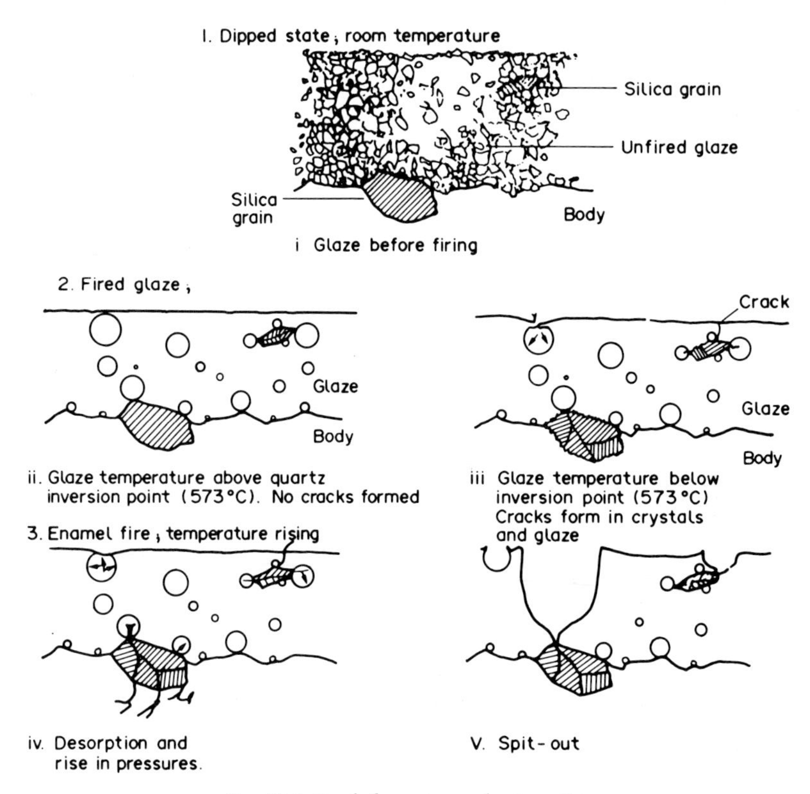

Fig. 12.4. Crack formation and spit-out[51]

a hole in the rear face of glost earthenware penetrating through the glaze layer and slighly into the body to provide an easy escape for the desorbing gases during the enamel fire. The single small hole is filled with a very soft, low melting point flux which does not markedly inhibit the venting of any gas from the body, yet which on cooling forms an impervious seal against the ingress of water and "soil" in service.

Excessive thicknesses of glaze should be avoided because any spit-out craters will be consequently large. The storage time of glost ware before entering the decorating line should be reduced to a minimum. Refiring "old" glost ware prevents spit-out from occurring. The decorating temperature should be chosen to be as low as possible consistent with achieving statutory requirements for metal release and for general durability.

12.14 "DROPPERS"

Although strictly a kiln problem the fault is seen as discoloured patches, roughly circular in shape, on the glaze surface. They are caused by drops of glassy material falling onto the ware from the roof of the kiln. At peak glost temperature, the fluxing components of the glaze volatilise and condense on cooler parts of the kiln, where they react with the refractory lining, to form a glass. In time, this deposit builds up to a sufficiently thick layer for the glass to flow and form drops. The colour of these drops is usually brownish from iron oxide contamination coming from the dissolved refractory.

When firing intermittently, the kiln can be examined regularly and any incipient droppers chipped away, but in continuous kilns preventive action must be taken. Tunnel kiln cars can be fitted with covering bats at the top of the setting to catch any falling droppers. In addition a coating of refractory wash can be applied to any kiln brickwork which is likely to become glazed. Many different refractory materials make suitable washes but recipes including powdered zircon are the most effective. The reaction products of zircon and volatilised fluxes are viscous and have high melting points; consequently flow is unlikely.

REFERENCES

1. Edwards, H. and Norris, A. W. *Trans. Brit. Ceram. Soc.* **56,** 133–153 (1957).
2. Franklin, C. E. L. *Trans. Brit. Ceram. Soc.* **64,** 549–566 (1965); *Trans. Brit. Ceram. Soc.* **65,** 277–287 (1966).
3. Dodd, A. E. *Dictionary of ceramics* Newnes (1964).
4. BS 1281:1966, Glazed ceramic tiles and tile fittings for internal walls.
5. BS 3402:1969, Quality of vitreous china sanitary appliances.
6. Mellor, J. W. *Trans. Brit. Ceram. Soc.* **34,** 1–112 (1935); *Trans. Brit. Ceram. Soc.* **36,** 443–465 (1937).
7. Mattyasovsky–Zsolnay, L. *J. Amer. Ceram. Soc.* **29,** 200–204 (1946).
8. Lehnhauser, W. The effect of humidity on moisture expansion of glazed ceramics. *Sprechasaal* **102** (23), 1061–1068 (1969).
9. Blakely, A. M. *J. Amer. Ceram. Soc.* **21,** 243–251 (1938).
10. Vaughan, F. and Dinsdale, A. *Trans. Brit. Ceram. Soc.* **61,** 1 (1962).
11. *The A. T. Green Book,* pp. 161–162. Pub British Ceramic Research Association (1959).
12. Smith, A. N. *Trans. Brit. Ceram. Soc.* **53,** 710–723 (1954).
13. Bullin, L. and Green, K. *Trans. Brit. Ceram. Soc.* **53,** 39–66 (1954).
14. *The A. T. Green Book*, pp. 162–163.
15. Johnson, R. *Trans. Brit. Ceram. Soc.* **75,** 1–5 (1976).
16. Harper, T. J. *Glass Technology* **3,** 171–175 (1962).
17. Scholze, H. *Eighth Int. Congress on Glass*, 69–83 (1968).
18. Williamson, W. O. *Trans. Brit. Ceram. Soc.* **59,** 455–477 (1960).
19. BS 3402:1969, Quality of vitreous china sanitary appliances.
20. Boulos, E. N. and Kreidl, N. J. *J. Canadian Ceram. Soc.* **41,** 83–91 (1972).
21. Bernal, J. D. and Fowler, R. H. *J. Chem. Phys.* **1,** 515–548 (1933).
22. Rawson, H. *Properties and Applications of Glass*, p. 189. Elsevier (1980).
23. Williams, J. P., Su, Y. S., Strzegowski, W. R., Butler, B. L., Hoover, H. L. and Altemose, V. O. *Bull. Amer. Ceram. Soc.* **55,** 524–527 (1976).
24. Communication from Dr J. K. Olby, Cookson Group Research Laboratories.
25. Adams, R. V. *Physics & Chemistry of Glasses,* **2,** 39–49 (1961).
26. Mattox, D. M. *J. Amer. Ceram. Soc.* **50,** 683 (1967).
27. Morey, G. W. *The Properties of Glass* Rheinhold (1938).
28. Dinsdale, A. *Trans. Brit. Ceram. Soc.* **62,** 321–388 (1963).

29. Shick, R. L. and Swarts, E. L. *J. Amer. Ceram. Soc.* **65,** 594–597 (1982).
30. Budworth, D. W., Smith, M. S. and Rose, M. E. *Trans. Brit. Ceram. Soc.* **73,** 57–59 (1974).
31. Morey, G. W. *The Properties of Glass*, p. 197. Rheinhold (1938).
32. Morey, G. W. *The Properties of Glass* p. 211 Rheinhold (1938).
33. BS 3402:1969.
34. Mattox, D. M. *J. Amer. Ceram. Soc.* **50,** 683 (1967).
35. BS 3402:1969.
36. Lepie, M. P. and Norton, F. H. *VII International Ceramic Congress* pp. 21–29. (1960).
37. McCartt, K. C. and Johnson, A. L. *Bull. Amer. Ceram. Soc.* **37,** 207–209 (1958).
38. BS 3402:1969, Quality of vitreous china sanitary ware.
39. Budworth, D. W. *Trans. Brit. Ceram. Soc.* **69,** 57–59 (1970).
40. "Zircosil in glazes". Published by Associated Lead Manufactures Ltd (1970). Text based on communication in *Trans. Brit. Ceram. Soc.* **58,** 532–564 (1959), by F. T. Booth and G. N. Peel.
41. Oldfield, L. F. and Wright, R. D. *Glass Tech.* **3,** 59–68 (1962).
42. Mellor, J. W. *Trans. Eng. Ceram. Soc.* **6,** 71–75 (1906–7).
43. Koenig, C. J. and Green, R. L. *Water Vapour in High Temperature Ceramic Processes*, p. 18. Ohio State University, USA.
44. Bailey, J. *J. Amer. Ceram. Soc.* **20,** 42–52 (1937).
45. Rose, M. E. and Wilshaw, T. R. *Trans. Brit. Ceram. Soc.* **74,** 171–175 (1975).
46. Hara, M. and Nakayama, J. *Glastechnische Berichte* **31,** 381–386 (1958).
47. Holland, A. J. and Turner, W. E. S. *J. Soc. Glass Tech.* **21,** 383–394 (1937).
48. Ashley, H. E. *Trans. Amer. Ceram. Soc.* **13,** 226–227 (1911).
49. Geller, R. F. and Creamer, A. S. *J. Amer. Ceram. Soc.* **14,** 624–630 (1931).
50. Wilkinson, W. T. and Dinsdale, A. *Trans. Brit. Ceram. Soc.* **60,** 33–66 (1961).
51. Wilkinson, W. T. and Dinsdale, A. *The A. T. Green Book*, pp. 255–268. BCRA (1959).

APPENDIX I

There are alternative ways of expressing the individual parts of a glaze or frit, the more common of which are:

1. Working recipes of raw material ingredients by weight.
2. Percentage composition of the elements expressed as oxides.
3. Molecular proportions of the element oxides.

Working Recipe

Working mill recipes take the form:

Frit	75
Felspar	15
China clay	10

Such recipes can be altered or modified with ease. Changing the working recipe might be obligatory, because normal raw material sources are lost or become uneconomic and others are substituted. A glaze can have the same ultimate formula or composition yet be constituted from different ingredients. For example, alumina and silica can be added as the single oxides, Al_2O_3 and SiO_2, or as china clay, $Al_2Si_2O_5(OH)_4$. Modifying the mill recipe might be necessary also to accommodate changes in the physical and chemical specification. Altering the quantity of one raw material alters the relationship between all components. An empirical approach to modification by changing the *working* recipe has a place in some areas of ceramics, but for an understanding of the subtle effects of these changes, the oxide compositions must be evaluated numerically.

A computer program can be prepared which converts working recipes into percentage terms and molecular "formulae".

Percentage Composition

The nature and quality of the constituent raw materials (obtained as a matter of routine control by chemical analysis) are readily available.

A felspar can be specified as:
Ideally: albite, soda felspar: $NaAlSi_3O_8$, which is equivalent to

SiO_2	68.7% (by weight)
Al_2O_3	19.5%
Na_2O	11.8%

The chemical analysis of a typical beneficiated mineral felspar (available in commercial quantities):

SiO_2	67.0% (by weight)
Al_2O_3	18.5%
Na_2O	5.5%
K_2O	8.5%
CaO	0.5%

Anhydrous sodium carbonate ("soda ash"), Na_2CO_3, has a molecular weight (mw) of 106. Our interest is in the components Na_2O (62) and CO_2 (44). These are present in sodium carbonate in weight percentage terms of:

$$Na_2O \quad \frac{62}{106} \times 100 = 58.5\%$$

This means that 58.5% of the weight of soda ash added to the frit batch is utilised within the structure of the final glassy frit:

$$CO_2 \quad \frac{44}{106} \times 100 = 41.5\%$$

When sodium carbonate is used in fritting, 41.5% of the added weight is lost as carbon dioxide.

The proportions of the relevant components of all commonly used ingredients of frit are easily accessible in published tables (cf Appendix IV), although these are often quoted for ideal (pure) compositions rather than the compositions of the technical/commercial grades of raw material normally available.

Percentage Composition from Batch Weight

The batch recipe for a simple leadless frit is:

Quartz sand	33.29 units
Powdered china clay	36.35 units
Ground dolomite	1.9 units
Ground limestone	10.28 units
Dehydrated borax	26.73 units
Soda ash	3.37 units

During the fritting of this mixture, combined water is lost from the china clay and carbon dioxide from the limestone, dolomite and soda ash. It is the consequent "glassified" oxides which can be expressed as a percentage composition. For the purposes of this example assume the raw materials are pure and use the fritting factors given in Appendix IV. Multiply the quantity of each ingredient given in the above example by the appropriate factors. Note that borax, dolomite and china clay contribute two oxides each.

Dehydrated borax	$26.73 \times 0.692\ B_2O_3$	$= 18.49\ B_2O_3$	
	$26.73 \times 0.308\ Na_2O$	$= 8.23\ Na_2O$	$= 10.20\ Na_2O$
Soda ash	$3.37 \times 0.585\ Na_2O$	$= 1.97\ Na_2O$	
Dolomite	$1.97 \times 0.208\ MgO$	$= 0.39\ MgO$	
	$1.97 \times 0.333\ CaO$	$= 0.63\ CaO$	$= 6.39\ CaO$
Limestone	$10.28 \times 0.56\ CaO$	$= 5.76\ CaO$	
China clay	$36.35 \times 0.395\ Al_2O_3$	$= 14.36\ Al_2O_3$	
	$36.35 \times 0.465\ SiO_2$	$= 16.90\ SiO_2$	$= 50.19\ SiO_2$
Quartz	33.29×1.0	$= 33.29\ SiO_2$	

Simplifying the above, we have:

SiO_2	50.19
Al_2O_3	14.36
CaO	6.39
MgO	0.39
Na_2O	10.20
B_2O_3	18.49
	100.02

Molecular Composition

The molecular "formulae" used are formalised expressions. Their existence on paper does not imply the existence of such molecular groupings in nature, but these Seger formulae are a convenient basis for comparing the property relationships of glazes and frits. With this method the balance of alkaline, intermediate and acidic oxides can be assessed. Some materials are assumed to be pure and calculations are often based on their theoretical formulae. Other materials are less pure and have appropriate factors for use in calculations. Two glazes—one including calcined alumina and silica and the other having china clay to provide the same amounts of silicon and aluminium in the glaze structure—can have similar calculated molecular formulae, yet would behave differently in the glost fire, perhaps because of minor or trace ingredients present in the clay but not in the purer prepared components. It is a question of experience to decide on the combination of raw materials to give the best fired glaze at an economical cost.

1. *To convert weight percentage compositions to molecular formulae, divide the percentages by the molecular weights of the oxides.*
2. *Sum the basic oxides.*
3. *Divide throughout by this sum.*

Example:

To convert a percentage composition to a molecular formula.

(i) Calculated compositions of frit as derived in previously:

SiO_2	50.20%
Al_2O_3	14.3(6)%
CaO	6.40%
MgO	0.40%
Na_2O	10.20%
B_2O_3	18.50%
	100.0(6)

(ii) Divide each percentage by the molecular weight of the oxide:

SiO_2	50.20÷ 60=0.837
Al_2O_3	14.36÷102=0.140
CaO	6.40÷ 56=0.114
MgO	0.40÷ 40=0.010
Na_2O	10.20÷ 62=0.164
B_2O_3	18.50÷ 70=0.264

(iii) It is a convention with frit and glaze for the sum of the basic oxides to equal unity. Here we have:

$$CaO + MgO + Na_2O$$
$$0.114+0.010+0.164=0.288$$

(iv) Therefore, divide each value in (ii) by 0.288:

SiO_2	$0.837 \div 0.288 = 2.91$	
Al_2O_3	$0.140 \div 0.288 = 0.49$	
CaO	$0.114 \div 0.288 = 0.395$	
MgO	$0.010 \div 0.288 = 0.035$	$=1.000$
Na_2O	$0.164 \div 0.288 = 0.570$	
B_2O_3	$0.264 \div 0.288 = 0.92$	

(v) These values can be put into a conventional framework:

CaO	0.395		$2.91 SiO_2$
MgO	0.035	$0.49\ Al_2O_3$	$0.92\ B_2O_3$
Na_2O	0.57		
	1.000		

Reconstitution of a Glaze Recipe from a Molecular Formula

There is a method using molecular proportions and reference should be made to a comprehensive treatment[1] for details.

The method here described reverses the procedure set out above.

Assume a glaze, firing about 1050° is represented by:

PbO	0.2459		SiO_2 2.844
Na_2O	0.2967	$0.334\ Al_2O_3$	
K_2O	0.0804		B_2O_3 0.426
CaO	0.3770		
	1.0000		

The presence of lead oxide in the analysis indicates the presence of a lead frit, either lead bisilicate or a lead borosilicate. Similarly, a borax frit will be the source of the B_2O_3 (rarely will the use of a boron mineral be either convenient or necessary). The compositions of the preferred source materials will be known. Some china clay will be necessary for the rheology and this will introduce both alumina and silica, (so we will need to apportion some alumina for clay to be a component) and felspar might be present to balance the requirements of alkalis. The recipe can, therefore take the form:

X	parts lead bisilicate frit
Y	parts borax frit
Z	parts china clay
W	parts sundries (felspar? quartz?)

To expand this formula, multiply the molecular part of each oxide in the molecular formula by its molecular weight:

PbO	$0.2459 \times 223 =$	54.836 parts PbO
Na_2O	$0.2967 \times 62 =$	18.352 parts Na_2O
K_2O	$0.0804 \times 94 =$	7.558 parts K_2O
CaO	$0.3770 \times 56 =$	21.112 parts CaO
Al_2O_3	$0.334 \times 102 =$	34.068 parts Al_2O_3
SiO_2	$2.844 \times 60 =$	170.64 parts SiO_2
B_2O_3	$0.426 \times 70 =$	29.82 parts B_2O_3

Lead oxide is added in the fully fritted form as lead bisilicate, the composition (wt%) of which is

PbO	64
SiO_2	33
Al_2O_3	3

Therefore, 54.836 parts PbO will be derived from

$$\frac{54.836}{64} \times 100 = 85.68 \text{ parts lead bisilicate}$$

which will at the same time introduce

85.68×0.33 parts of SiO_2 $= 28.27$ SiO_2
85.68×0.03 parts of $Al_2O_3 =$ 2.57 Al_2O_3

All the B_2O_3 must be introduced as a frit. A suitable low melting point frit has the composition (wt%).

SiO_2	52.77
Al_2O_3	5.52
CaO	12.98
Na_2O	9.1
K_2O	1.09
B_2O_3	18.54

Therefore, 29.82 parts B_2O_3 will be introduced by

$$\frac{29.82}{18.54} \times 100 = 160.84 \text{ parts borax frit.}$$

which quantity of frit will also contribute

SiO_2	$160.84 \times 0.5277 = 84.8$ parts SiO_2
Al_2O_3	$160.84 \times 0.0552 = 8.88$ parts Al_2O_3
CaO	$160.84 \times 0.1298 = 20.88$ parts CaO
Na_2O	$160.84 \times 0.091 = 14.64$ parts Na_2O
K_2O	$160.84 \times 0.0109 = 1.75$ parts K_2O

We have added thus far

K_2O 1.75 parts; Na_2O 14.64 parts

but the composition demands 18.35 Na_2O and 7.56 K_2O.

The balance of $(18.35-14.64)$ $Na_2O = 3.71$ Na_2O
and $(7.56-1.75)$ $K_2O = 5.81$ K_2O

can be derived from felspar.

A suitable felspar has the composition (wt%)

SiO_2	67
Al_2O_3	18.5
K_2O	8.5
Na_2O	5.5
(CaO, Fe_2O_3 ect.)*	0.5

*For simplicity these small components are not considered in the calculations.

5.81 K_2O is derived from $\frac{5.81}{0.085} = 68.35$ felspar from which we obtain

$$68.35 \times 0.055\ Na_2O = 3.76\ Na_2O^*$$

and $68.35 \times 0.185\ Al_2O_3 = 12.64\ Al_2O_3$

plus $68.35 \times 0.67\ SiO_2 = 45.79\ SiO_2$

*The requirements for Na_2O is therefore fully satisfied.

We have added thus far in three components

Al_2O_3 $12.64 + 8.88 + 2.57 = 24.09\ Al_2O_3$

SiO_2 $45.79 + 84.87 + 28.27 = 158.93\ SiO_2$

but the composition requires 34.068 Al_2O_3 and 170.64 SiO_2

The balance of $(34.07 - 24.09)\ Al_2O_3 = 9.98\ Al_2O_3$

and $(170.64 - 158.93)\ SiO_2 = 11.71\ SiO_2$

The raw materials so far selected would not themselves give a workable glaze. For the rheology and working (dipping or spraying) qualities and to provide adequate suspension properties for the other components, the glaze will need a proportion of china clay in its make-up.

Ideally china clay contains say

Al_2O_3	39.5% (wt)
SiO_2	46.5% (wt)

Considering the requirement for Al_2O_3 we can, by adding china clay, obtain 9.98 parts Al_2O_3 from

$$\frac{9.98}{0.395} = 25.3 \text{ china clay}$$

This quantity of clay contributes $25.3 \times 0.465 = 11.74\ SiO_2$ and thus completing the requirements for all oxides.

The recipe, therefore, becomes

18.7 parts lead bisilicate

160.8 parts borax frit

68.4 parts potash felspar

25.3 parts china clay

REFERENCE

1. Griffiths R. & Radford C. "Calculations in Ceramics" (Maclaren) (1965).

APPENDIX II

CONVERSION OF DRY GLAZE RECIPE TO SLIP RECIPE

A sample glaze recipe in dry weight is

Ground leadless frit	75 parts
Powdered china clay	10 parts
Ground potash felspar	15 parts

It is needed in slop form at 29 oz to the pint, equivalent to a slip density of 1.45 kg l^{-1} (g ml^{-1}) or specific gravity of 1.45. The specific gravity of the dry glaze is, say 2.5, an average value typical for leadless materials. The ratio of dry materials:water can be determined using Brongniart's formula.[1,2]

Brongniart's formula:

Imperial units $W=(P-20)\dfrac{g}{g-1}$ Metric units $M=(D-1)\dfrac{g}{g-1}$

where W is the weight of dry glaze in ounces per pint of slip,
M is the weight of dry glaze per litre of slip in kg.
P is the slip density in oz pt^{-1},
D is the slip density in kg l^{-1}
g is the specific gravity of the dry solid material.

Imperial Units

Dry weight in 1 pint is $\dfrac{(29\text{–}20)\times 2.5}{(2.5-1)}$, i.e. 15 oz dry content

Therefore, for 1 gallon we require $8\times 15=120$ oz$=7.5$ lb dry glaze.

To make 1 gallon of slop leadless glaze at 29 oz pt^{-1} we need

$$(8\times 29)-120=112 \text{ oz} =5.6 \text{ pt water}$$

Therefore, batch recipe is 5.625 lb frit, 0.75 lb clay, 1.125 lb felspar } 7.5 lb
plus 5.6 pt water.

Metric Units

$$\text{Dry weight in 1 litre} \frac{(1450-1000)\times 2.5}{(2.5-1)}, \text{ i.e. 750 g dry content}$$

Therefore, for 1 litre we require 750 g dry glaze.
For 5 litres of glaze we need 3750 g glaze.
For 5 litres at 1.45 s.g. we need

$$5\times(1450-750)=3500 \text{ g water.}$$

Therefore, batch recipe is 2812.5 g frit, 375 g clay, 562.5 g felspar } 3750 g
plus 3500 g water

Note that with leadless glazes approximately equal weights of dry glaze and water give a workable glaze suspension. With low solubility (lead) glazes the basic ratio is approximately 1.6 parts dry glaze to one part water.

REFERENCES

1. Brongniart, A. *Traite des Arts Ceràmiques* **1,** p. 249 (1854)
2. Dodd, A. E. *Dictionary of Ceramics* p. 39 Newnes (1964).

APPENDIX III

FLUID MIXINGS

Dry Content

Some compositional adjustments can be made in the fluid state by mixing measured volumes of glaze slips having known fluid densities. Reference is still made to the slip density as "pint weight", but increasingly the metric value is reported. Standard imperial and metric measures are available.

The relationship between a definite volume of slip and the quantity of solid matter in suspension can be determined. For some application methods it might be necessary to adjust a high density slip to one of a lower density. The relationship noted in Appendix II can be used,

Imperial units (pint weight)

$$\text{Dry content } W = \frac{(P-20)\,g}{(g-1)}$$

where W is weight in ounces of dry material in 1 pint of slip,
P is weight in ounces of 1 pint volume,
g is specific gravity of glaze when dry.

Metric units (slip density)

$$\text{Dry content } W = \frac{(L-1000)\,g}{(g-1)}$$

where W is weight in grams of dry material in 1 litre of slip,
L is weight in grams of 1 litre slip volume,
g is specific gravity of glaze when dry.

Thus the quantity of dry material (Sg 2.5) in a leadless glaze slip with pint weight of 30 oz per pint, fluid density 1.5 can be found.

Imperial units.

$$\text{Dry content } W = \frac{(30-20)\times 2.5}{(2.5-1)}$$

$$= \frac{10\times 2.5}{1.5} = 16\ 2/3 \text{ oz}$$

Metric units

$$\text{Dry content } W = \frac{(1500-1000)\times 2.5}{(2.5-1)}$$

$$= \frac{500\times 2.5}{1.5} = 833.3\text{g}$$

If a reduction in the fluid density of a stock glaze is needed, the formulae in Appendix II can be reduced to:

$$\text{Imperial units } V = \frac{S-D}{(D-20)}$$

where S=pint weight of stock,
D=desired pint weight,
V=volume in pints of water to be added.

$$\text{Metric units } V = \frac{S-D}{(D-1000)}$$

where S=slip density in grams per litre,
D=desired slip density,
V=volume in mls of added water.

REFERENCES

1. Brongniart, A. *Traite des Arts Ceràmiques* **1,** p. 249 (1854)
2. Dodd, A. E. *Dictionary of Ceramics* p. 39 Newnes (1964).

APPENDIX IV
CONVERSION FACTORS

Material	Approx. molecular weight	Theoretical formula	Oxides supplied	Factor	Reciprocal
Alumina	102	Al_2O_3	Al_2O_3	1.000	1.000
Alumina hydrate	78	$Al(OH)_3$	Al_2O_3	0.654	1.531
Ammonium fluozirconate	241	$(NH_4)_2\ ZrF_6$	ZrO_2	0.51	1.965
			ZrF_4	0.693	1.442
			F	0.473	2.114
Ammonium di-hydrogen phosphate	115	$NH_4H_2PO_4$	P_2O_5	0.617	1.620
Antimony oxide	292	Sb_2O_3			
Barium carbonate	197	$BaCO_3$	BaO	0.777	1.288
Barium sulphate	233	$BaSO_4$	BaO	0.657	1.523
Borax, crystalline	381	$Na_2B_4O_7 10H_2O$	B_2O_3	0.365	2.738
			Na_2O	0.163	6.135
			$Na_2B_4O_7$	0.528	1.895
			H_3BO_3	0.649	1.542
Borax, anhydrous	202	$Na_2B_4O_7$	Na_2O	0.308	3.246
			B_2O_3	0.692	1.445
Boric acid	62	H_3BO_3	B_2O_3	0.563	1.776
Borocalcite	303	$CaO.2B_2O_3.6H_2O$	B_2O_3	0.438	2.283
			CaO	0.279	3.584
Cadmium carbonate	172	$CdCO_3$	CdO	0.746	1.340
Calcium carbonate (chalk, lime, whiting, marble, calcite)	100	$CaCO_3$	CaO	0.560	1.784
Calcium phosphate (bone ash)	310	$Ca_3(PO_4)_2$	CaO	0.542	1.844
			P_2O_5	0.458	2.184
Calcium sulphate (anhydrite)	136	$CaSO_4$	CaO	0.412	2.428
Calcium sulphate (gypsum)	172	$CaSO_4.2H_2O$	CaO	0.326	3.067
Cryolite	210	Na_3A1F_6	Na_2O	0.443	2.258
			Al_2O_3	0.243	4.118
			F	0.544	1.838

Material		Formula	Oxide		
Dolomite	184	$MgCO_3.CaCO_3$	MgO	0.201	4.975
			CaO	0.332	3.012

Felspar

	Theoretical Na spar	Theoretical K spar	Theoretical Ca spar
Na_2O	.118	—	—
Al_2O_3	.195	.183	.367
SiO_2	.687	.647	.432
K_2O	—	.169	—
CaO	—	—	-.201

	FFF	HN	FM59	Four Crowns	NCNS
K_2O	8.5	10.2	9.3	4.4	9.2
Na_2O	5.5	3.2	2.8	3.6	8.1
Al_2O_3	18.5	18.5	17.0	13.7	24.75
SiO_2	67.0	66.7	69.4	76.7	56

Material		Formula	Oxide		
Ferric hydroxide	107	$Fe(OH)_3$	Fe_2O_3	0.747	1.339
Fluorspar	78	CaF_2	CaO	0.718	1.392
			F	0.487	2.052
Hydrated lime	74	$Ca(OH)_2$	CaO	0.756	1.321
Iron oxide (ferric)	160	Fe_2O_3	Fe_2O_3	1.000	1.000
Kaolin	258	$Al_2O_3.2SiO_2.2H_2O$	Al_2O_3	0.395	2.532
			SiO_2	0.465	2.150
Kyanite	162	$Al_2O_3.SiO_2$	Al_2O_3	0.630	1.587
			SiO_2	0.370	2.703
Lead bisilicate (usually contains some alumina)	343	$PbO.2SiO_2.xAl_2O_3$	PbO	0.640	1.563
			SiO_2	0.330	3.030
			Al_2O_3	0.030	33.333
Lead chromate	323	$PbCrO_4$	PbO	0.690	1.449
			Cr_2O_3	0.235	4.225
Lead fluoride	245	PbF_2	PbO	0.95	1.052
			F	0.155	6.450
Litharge	223	PbO	PbO	1.000	1.000
Lithium carbonate	74	Li_2CO_3	Li_2O	0.406	2.463
Limestone	100	$CaCO_3$	CaO	0.560	1.784
Magnesium carbonate	84	$MgCO_3$	MgO	0.478	2.091
Magnesium carbonate basic	382	$3MgCO_3.Mg(OH)_2.4H_2O$	MgO	0.419	2.387
Magnesium oxide	40	MgO	MgO	1.000	1.000
Magnesium sulphate	246	$MgSO_4.7H_2O$	MgO	0.161	6.211
			SO_2	0.258	3.876
			SO_3	0.320	3.125
Manganese carbonate	115	$MnCO_3$	MnO	0.617	1.621

				many grades available	
Manganese dioxide, pyrolusite	87	MnO_2	MnO	0.816	1.225
Manganese nitrate	179	$Mn(NO_3)_2$	MnO	0.397	2.519
Manganese sulphate	151	$MnSO_4$	MnO	0.469	2.132
Pearl ash (potassium carbonate)	165	$K_2CO_3.1\frac{1}{2}H_2O$	K_2O	0.570	1.754
Potassium carbonate (calcined)	138	K_2CO_3	K_2O	0.682	1.466
Potassium dichromate	294	$K_2Cr_2O_7$	K_2O	0.32	3.125
			Cr_2O_3	0.52	1.923
Potassium fluozirconate	283	K_2ZrF_6	K_2O	0.333	3.000
			ZrO_2	0.435	2.300
			F	0.403	2.480
			ZrF_4	0.590	1.694
Potassium nitrate	101	KNO_3	K_2O	0.466	2.145
Red lead	686	Pb_3O_4	PbO	0.977	1.024
Saltcake	142	Na_2SO_4	Na_2O	0.437	2.290
			SO_3	0.563	1.776
Sodium carbonate (soda ash)	106	Na_2CO_3	Na_2O	0.585	1.710
Sodium fluoride	42	NaF	Na_2O	0.739	1.356
			F	0.453	2.207
Sodium chloride	58.5	NaCl	Na_2O	0.530	1.887
Sodium nitrate	85	$NaNO_3$	Na_2O	0.365	2.742
Sodium phosphate	322	$Na_2HPO_4.10H_2O$	Na_2O	0.193	5.192
			P_2O_5	0.221	4.535
Sodium silicate (Neutral Glass)		$Na_2O.xSiO_2$	Na_2O	0.227	4.396
			SiO_2	0.768	1.302
Sodium silicofluoride	188	Na_2SiF_6	Na_2O	0.330	3.030
			SiO_2	0.319	3.133
			F	0.606	1.650
Stone	580		SiO_2		72.4
			Al_2O_3		16.32
			Fe_2O_3		
			TiO_2		
			MgO		
			CaO		1.55
			K_2O		4.66
			Na_2O		3.50

Strontium carbonate	148	$SrCO_3$	SrO	0.702	1.425
Talc	379	$3MgO.4SiO_2H_2O$	MgO	0.319	3.135
			SiO_2	0.634	1.577
Tin oxide (stannic oxide)	151	SnO_2	SnO_2	1.000	1.000
Titanium dioxide	810	TiO_2	TiO_2	1.000	1.000
Ulexite (sodium calcium borate)	810	$Na_2O.2CaO.5B_2O_3.16H_2O$	B_2O_3	0.429	2.331
			NaO	0.077	12.987
			CaO	0.138	7.042
Zinc oxide	81	ZnO	ZnO	1.000	1.000
Zircon ("Zircosil")	183	$ZrSiO_4$	ZrO_2	0.670	1.492
			SiO_2	0.330	3.030
Zirconia	123	ZrO_2	ZrO_2	1.000	1.000
Zirconium tetrafluoride	167	ZrF_4	ZrO_2	0.737	1.358
			F	0.455	2.198

Information about less common minerals used in glazes and frits is in the text.

Appendix V
Glaze Stains

Colour		Stain type	Maximum firing temperature	Desirable glaze characteristics for optimum effect
Blue,	dark	Olivine, cobalt silicon	1400°C	Suitable for use in all glazes
	light	Spinel, cobalt zinc aluminium	1300°C	Suitable for use in all glazes
	light	Zircon, vanadium	1300°C	Suitable for use in all glazes
Green,	dark	Spinel, chromium cobalt (zinc)	1400°C	Suitable for use in most zinc-free glazes
	light	Garnet, calcium chromium silicon	1200°C	Zinc-free, high calcium compositions
	light	Zircon, vanadium/praseodymium	1250°C	Suitable for use in all glazes
Orange		Pyrochlore, lead antimony	1050°C	High lead content glazes
		Rutile, titanium chromium	1100°C	Leadless glazes
		Baddeleyite (indium, yttrium) vanadium	1400°C	Suitable for use in all glazes, especially those firing at high temperatures
Yellow		Zircon, praseodymium	1250°C	Suitable for use in all glazes
		Cassiterite, vanadium	1300°C	Suitable for use in all glazes
		Cadmium sulphide	1050°C	Glaze must be specially designed to give temperature stability to the pigment
Red		Cadmium sulphoselenide	1050°C	Glaze must be specially designed to give temperature stability to the pigment
Pink		Zircon, cadmium sulphoselenide	1300°C	Suitable for use in all glazes
		Zircon, iron	1300°C	Suitable for use in all glazes
		Sphene, chromium tin silicon	1200°C	Zinc-free, high calcium compositions
		Spinel, chromium aluminium zinc	1300°C	High zinc and aluminium glaze with low calcium content
Lilac		Cassiterite, chromium	1200°C	Zinc-free high calcium compositions
Brown,	dark	Spinel, iron chromium (nickel)	1300°C	Very suitable for use in zinc-containing glazes
		iron chromium zinc	1300°C	
	light	iron aluminium chromium zinc	1300°C	
Black		Spinel, chromium cobalt iron nickel	1300°C	Zinc-free compositions
Grey		Cassiterite, antimony vanadium	1300°C	Suitable for use in all glazes

Appendix VI

Throughout technical literature on glaze and glaze technology an enormous number of frit compositions have been reported and their properties discussed. Many of these data are spurious. The formulations here recorded (and which are typical examples of present practice) are of proven value and are in regular production for use in commercial glazes. They are reproduced by kind permission of Cookson Group plc.

VI.1 LEAD FRITS

Softening point	% Linear expansion at 500°C	Specific gravity	Refractive index	% Lead solubility (ground 3000 $cm^2\ g^{-1}$)	Molecular weight	Molecular formula	Na_2O	K_2O	PbO	Al_2O_3	SiO_2	TiO_2	Notes
650°C	0.41	5.00	>1.8	<4.0	325.4	PbO } 0.118 TiO_2 { 1.54 SiO_2	—	—	68.7	—	28.5	2.9	Titania based high lead silicate. Low solubility for matt & crystalline glazes.
690°C	0.36	4.75	1.75	<4.0	333.7	PbO } 0.148 Al_2O_3 { 1.62 SiO_2	—	—	66.5	4.5	29.0	—	Low solubility frit with increased lead content for brighter glazes.
700°C	0.32	4.57	1.73	<3.5	347.5	PbO } 0.042 Al_2O_3 { 2.00 SiO_2	—	—	64.2	1.2	34.6	—	Low solubility frit for tableware and wall tile glazes.
700°C	0.28	4.55	1.73	<3.5	353.1	PbO } 0.142 Al_2O_3 0.182 TiO_2 { 1.68 SiO_2	—	—	63.0	4.2	28.6	4.2	Alumina/titania based lead silicate for textured glazes.
700°C	0.33	4.60	—	<2.5	346.2	PbO } 0.095 TiO_2 { 1.93 SiO_2	—	—	64.4	—	33.4	2.2	Titania based lead silicate for matt and vellum glazes.
710°C	0.34	4.57	1.73	<3.5	343.5	PbO } 0.086 Al_2O_3 { 1.86 SiO_2	—	—	64.9	2.6	32.5	—	Standard lead bisilicate for all types of glazes.
725°C	0.32	4.45	1.71	<3.5	347.3	0.924 PbO, 0.020 Na_2O, 0.056 K_2O } 0.087 Al_2O_3 { 2.05 SiO_2	0.4	1.4	59.4	2.5	35.4	—	General purpose lead frit for low solubility glazes
750°C	0.27	4.57	1.73	<3.0	366.2	PbO } 0.247 Al_2O_3 { 1.97 SiO_2	—	—	61.0	7.0	32.0	—	High alumina lead silicate for all types of glazes Good craze resistance.

VI.2 Lead borosilicate frits

Softening point	% Linear expansion at 500°C	Specific gravity	Refractive index	% Lead solubility (ground 3000 $cm^2\ g^{-1}$)	Molecular weight	Molecular formula	Na_2O	K_2O	CaO	PbO	Al_2O_3	B_2O_3	SiO_2	Notes
880°C	0.24	4.01	1.54	<2.0	383.0	0.362 PbO, 0.188 Na_2O, 0.375 CaO, 0.075 K_2O : 0.207 Al_2O_3 : 3.78 SiO_2, 0.20 B_2O_3	3.1	1.8	5.5	21.0	5.5	3.8	59.4	Low expansion. Low solubility.
880°C	0.27	4.03	1.55	<3.0	354.5	0.202 Na_2O, 0.391 PbO, 0.407 CaO : 0.168 Al_2O_3 : 3.10 SiO_2, 0.40 B_2O_3	3.5	—	6.4	25.0	4.8	7.4	52.6	Low solubility. Good colour development.

VI.3 Zircon opaque frits

Softening point	% Linear expansion at 500°C	Specific gravity	Refractive index	Molecular weight	Molecular formula	Na_2O	K_2O	CaO	PbO	ZnO	ZrO_2	Al_2O_3	SiO_2	B_2O_3	Notes
820°C	0.27	4.32	1.54	380.0	0.140 Na_2O, 0.080 K_2O, 0.310 CaO, 0.280 PbO, 0.210 ZnO : 0.220 Al_2O_3 : 2.82 SiO_2, 0.51 B_2O_3, 0.35 ZrO_2	2.3	1.7	4.6	16.3	4.2	11.3	5.9	44.4	9.3	Low solubility zircon frit for opaque glazes.
820°C	0.28	3.08	1.51	390.0	0.390 Na_2O, 0.110 ZnO, 0.500 CaO : 0.165 Al_2O_3 : 3.50 SiO_2, 0.82 B_2O_3, 0.30 ZrO_2	6.0	—	7.4	—	2.3	9.7	4.5	55.0	14.7	Leadless Zircon frit for extra white glazes.
850°C	0.27	3.10	1.52	392.5	0.441 CaO, 0.258 Na_2O, 0.185 K_2O, 0.141 ZnO : 0.316 Al_2O_3 : 3.52 SiO_2, 0.61 B_2O_3, 0.25 ZrO_2	4.1	4.4	6.3	—	2.9	7.9	8.2	55.2	11.0	Leadless Zircon frit for extra white glazes.
850°C	0.31	2.82	—	249.5	0.260 Na_2O, 0.063 K_2O, 0.153 CaO, 0.524 ZnO : 0.120 Al_2O_3 : 2.00 SiO_2, 0.45 B_2O_3, 0.12 ZrO_2	6.4	2.4	3.4	—	17.0	5.7	4.5	48.1	12.5	Zinc/zircon matt for tile and tableware glazes.

Index